AN INTRODUCTION TO GNSS
A primer in using Global Navigation Satellite Systems for positioning and autonomy
Third Edition

Published by NovAtel Inc.
Hexagon Calgary Campus
10921 14th Street NE
Calgary, Alberta
T3K 2L5
Canada
hexagon.com/autonomy-and-positioning

This document and the information contained herein are the exclusive properties of Antcom Corporation, Autonomous Stuff LLC, NovAtel Inc., Veripos Limited, and/or their affiliates within the Hexagon Autonomy & Positioning division ("A&P").

No part of this document may be reproduced, displayed, distributed, or used in any medium, in connection with any other materials, or for any purpose without prior written permission from A&P. Applications for permission may be directed to communications.nov.ap@hexagon.com. Unauthorised reproduction, display, distribution or use may result in civil as well as criminal sanctions under the applicable laws. A&P aggressively protects and enforces its intellectual property rights to the fullest extent allowed by law.

This document and the information contained herein are provided AS IS and without any representation or warranty of any kind. A&P disclaims all warranties, express or implied, including but not limited to any warranties of merchantability, non-infringement and fitness for a particular purpose. Nothing herein constitutes a binding obligation on A&P.

The information contained herein is subject to change without notice.

© 2015-2023. All rights reserved. A list of entities within the Hexagon Autonomy & Positioning division is available at hexagonpositioning.com.

ISBN: 978-0-9813754-1-0

> "Their 'we-do-it-not-because-it-is-easy-but-because-it-is-hard' attitude inspired me."
>
> —Dr. Robert Thirsk, Astronaut, doctor, engineer

FOREWORD

The urge to explore is innate — and space presents humanity with a boundless frontier to be explored. For myself, the notion of space exploration was planted in my mind when I was a child. Growing up in the 1960s, I followed the exploits of the early astronauts who boldly went where no one had gone before. What courage!

I watched in awe as Armstrong and Aldrin guided their spidery lunar module down to a pin-point landing on the Moon and then, a few hours later, bounded about its magnificently desolate surface. Their "we-do-it-not-because-it-is-easy-but-because-it-is-hard" attitude inspired me. In later years, I jumped through all the academic and career hoops that might someday help me realise my own dream of spaceflight.

Beyond this self-indulgent motivation to explore the great unknown, the development of space over the past decades has brought about pragmatic and untold benefits to society at large. We take for granted that many everyday services are enabled by space technology — how we communicate, how we know whether to grab an umbrella as we step out the door in the morning, even how we're entertained. Space even affords a unique laboratory setting to advance research in ways that aren't possible on Earth.

During both my forays into space, I relied upon positioning and automation technologies to complete my missions. During ascent of the shuttle, for example, data from the vehicle's three inertial measurement units were fed to the flight software steering the engine gimbals so that we arrived in space at a targeted position and velocity.

On orbit, I worked with science teams on the ground to conduct a host of international investigations — from medical science to fluid physics to robotics. Whether crystalising a large protein molecule associated with a debilitating congenital illness — impossible to do on Earth due to gravity — to testing control algorithms for satellite systems using free-flying robots, every experiment provided new insights. Knowing that the data we collected would be applied to social needs on the ground brought a feeling of satisfaction and gratitude for the opportunity to serve. And the view out the spacecraft window wasn't too bad either!

What unfolds over the next decades will be even more exciting. We will watch in wonder as the next generation of astronauts ventures farther into the solar system.

Enterprising industrialists will develop novel means to harness resources from the Sun and near-Earth asteroids to support the needs of civilization. Enhancements of existing space technologies will improve sustainability of life on Earth. In particular, new applications of autonomy and positioning technologies will transform how we grow our food, monitor the environment, manage our natural resources and transport goods about the global supply chain.

Which brings me to the subject of this book. Global Navigation Satellite Systems

Aurora borealis over Alberta, Canada, taken from the International Space Station

(GNSS) are at the heart of answering the questions: where am I and how do I get to my destination? The answers lie in precision positioning, navigation and timing (PNT) capabilities enabled by GNSS. The accuracy and speed of the technologies detailed in this book have advanced to the point where cars, ships, planes and spacecraft can know their locations in real-time with unprecedented precision. Their ready adoption across applications like mining and autonomous vehicles has been enabled by the integration of advanced sensors and increasingly sophisticated positioning solutions.

What does this have to do with sustainability? Greater efficiencies and reduced costs are achieved through autonomous operations, from self-driving transport trucks that shave hours off shipping times and enable unmanned ore transport in mining operations, to agricultural positioning technologies that ensure year-after-year repeatability in crop production.

Although I'm no longer an active astronaut, my passion for innovation guides my continuing mission to make our world a better place. One way I do that is by encouraging today's young explorers to work outside of their comfort zones, to participate in collaborative, team-oriented ventures and to pursue audacious dreams. The realisation of such dreams will only be possible when built upon a foundation of lifelong learning and advanced skills.

If you're reading this book, you've already launched on the right trajectory to this bright and sustainable future!

TABLE OF CONTENTS

1 Overview — 03
2 Basic concepts — 11
3 GNSS constellations — 27
4 GNSS error sources — 43
5 Resolving errors — 47
6 Sensor fusion — 61
7 GNSS threats — 67
8 Autonomy — 73
9 GNSS applications and equipment — 83

Appendix A – Acronyms — 100
Appendix B – GNSS glossary — 102
Appendix C – Standards and references — 108
Appendix D – Acknowledgements — 109

GNSS overview

> "New ideas pass through three periods: 1) It can't be done. 2) It probably can be done, but it's not worth doing. 3) I knew it was a good idea all along!"
>
> –**Arthur C. Clarke**, British author, inventor and futurist.

Most of us now know that GNSS "was a good idea all along" and that we are well into Clarke's third phase.

The basic concepts of satellite positioning are very easy to understand. They are so straightforward, in fact, that one of our employees was asked by his daughter to explain it to her grade 4 class.

Before the class started, he set up the following demonstration. He tacked cardboard figures of three satellites to the walls and ceiling of the classroom, as shown in **Figure 1**. Each "satellite" had a length of string stapled to it. He marked a location on the floor with a movable dot, then drew the strings down and marked where they all reached the dot. The strings now represented the distances from the dot to the individual satellites. He recorded the location of the dot and

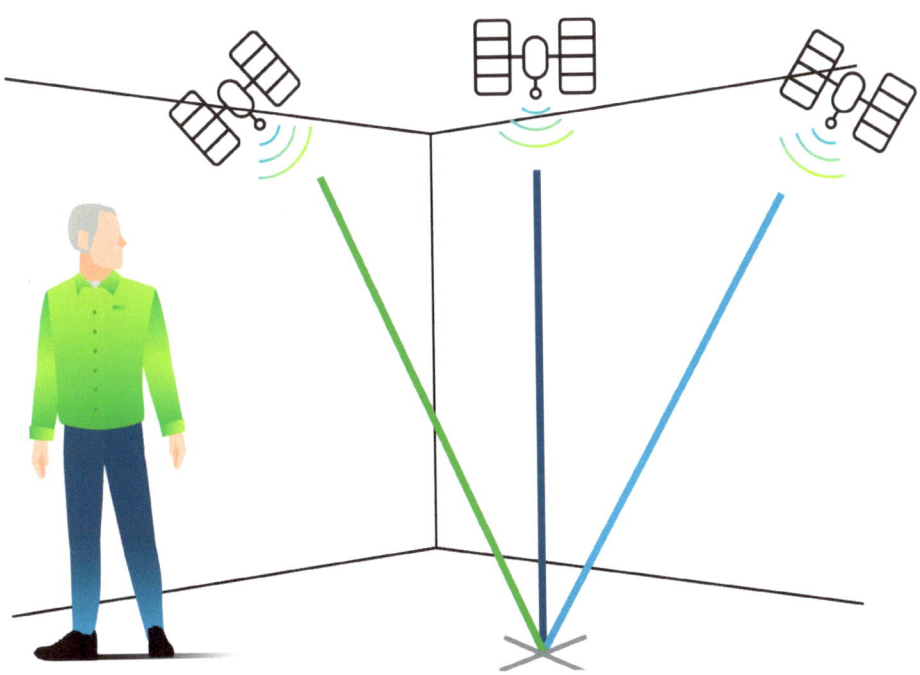

GNSS OVERVIEW

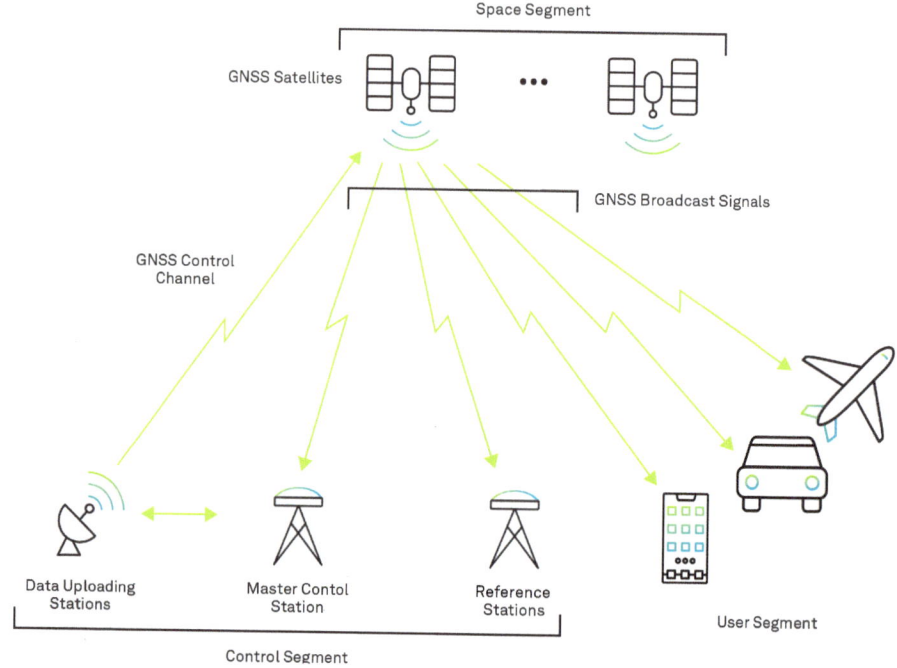

Figure 2 GNSS segments

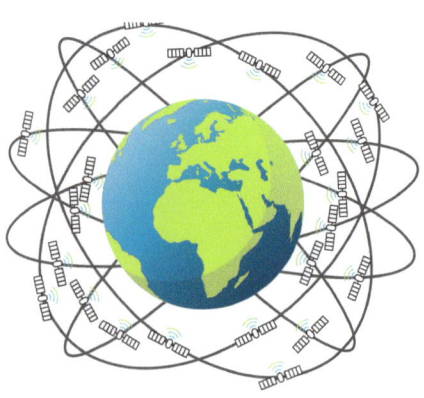

Figure 3 GNSS satellite orbits

When the students came into the classroom, our employee had them use the strings to determine the location. To do this, the students drew the strings down until the ends came together at one point on the floor. They marked this point with a movable dot and compared it with the previously marked position. They were very close. This simple demonstration showed that if you know the location of three satellites and your distance from them, you can determine your position.

The determination of position is made quite a bit more complicated by several factors: the satellites are moving, the signals from the satellites are very weak by the time they reach the Earth, the atmosphere interferes with the transmission of radio signals and, for cost reasons, the user equipment is not as sophisticated as the equipment in the satellites.

GNSS OVERVIEW

> "The more you explain it, the more I don't understand it."
>
> –**Mark Twain**, American author and humourist.

We agree with Mark Twain. We will provide a more detailed explanation of position determination in Chapter 2.

GNSS constellations

Although you may already be familiar with the term "GPS" (Global Positioning System), you may not have heard the term "GNSS" (Global Navigation Satellite System), which is used to describe the collection of satellite positioning systems, called "constellations," from various countries that are in operation or planned.

GPS (United States)

GPS was the first GNSS system and was launched in the late 1970s by the U.S. Department of Defense. The GPS constellation provides global coverage.

GLONASS (Russia)

GLONASS is operated by the Russian government. The GLONASS constellation provides global coverage.

Galileo (European Union)

Galileo is a civil GNSS system operated by the European Global Navigation Satellite Systems Agency (GSA). Galileo provides global coverage.

BeiDou (China)

BeiDou is the Chinese navigation satellite system. The BeiDou system provides global coverage with enhanced service for the China region.

NavIC (India)

The Navigation with Indian Constellation (NavIC) system provides service to India and the surrounding area.

QZSS (Japan)

The Quasi-Zenith Satellite System (QZSS) is a regional navigation satellite system that provides service to Japan and the Asia-Oceania region.

In Chapter 3, we will provide additional information about these systems.

GNSS architecture

> "The future ain't what it used to be."
>
> –**Yogi Berra**, former Major League Baseball player and manager.

Mr. Berra is correct. The implementation of GNSS has really changed things.

GNSS consist of three major components or "segments:" the space segment, the control segment and the user segment. These are illustrated in **Figure 2**.

Space segment

The space segment consists of GNSS satellites orbiting 19,000 to 36,000 kilometres (11,800 to 22,400 miles) above the Earth. Each GNSS has its own constellation of satellites, arranged in orbits to provide the desired coverage, as illustrated in **Figure 3**.

Each satellite in a GNSS constellation broadcasts a signal identifying itself and providing its precise time, orbit location and system health status. To illustrate, consider the following. You are downtown. You call a friend. Your friend is not at home, so you leave a message:

> *This is your friend [identity]. The time is 1:35 p.m. [time]. I am at the northwest corner of 1st Avenue and 2nd Street and I am heading towards your place [orbit]. I am OK, but I am a bit thirsty [status].*

Your friend returns a couple of minutes later, listens to your message and "processes" it, then calls you back and

An Introduction to GNSS, Third Edition   5

GNSS OVERVIEW

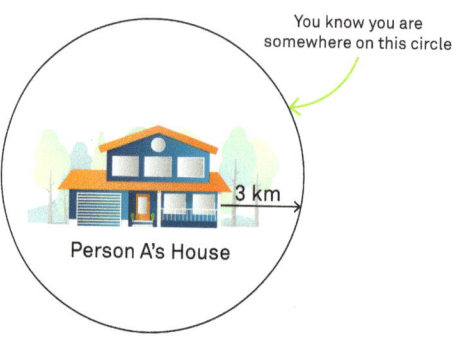

Figure 4 Illustration of trilateration— knowing one distance

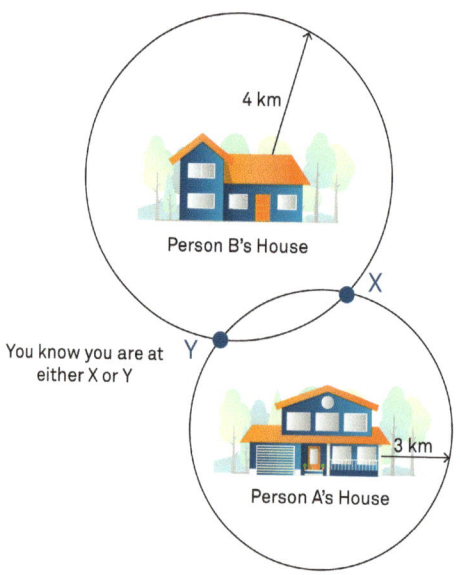

Figure 5 Illustration of trilateration— knowing two distances

suggests you come up a slightly different way. Effectively, your friend has given you an "orbit correction."

Control segment

The control segment comprises a ground-based network of master control stations as well as data uploading stations and monitor stations. In the case of GPS, the control segment includes a master control station, an alternate master control station, 11 command and control antennas and 16 monitoring sites located throughout the world.

In each GNSS system, the master control station adjusts the satellites' orbit parameters and onboard high-precision clocks when necessary to maintain accuracy.

Monitor stations, usually installed over a broad geographic area, monitor the satellites' signals and status and relay this information to the master control station. The master control station analyses the signals and transmits orbit and time corrections to the satellites through data uploading stations.

User segment

The user segment consists of equipment that processes the received signals from the GNSS satellites and uses them to derive and apply location and time information. The equipment ranges from smartphones and handheld receivers used, for example, by hikers, to sophisticated, specialised receivers used for high-end survey and mapping applications.

GNSS signals

GNSS radio signals are quite complex. Their frequencies range from 1.17 to 1.61 GHz (gigahertz), or 1.17 to 1.61 billion cycles per second. For comparison, GNSS operates at frequencies that are higher than FM radio but lower than a microwave oven. By the time GNSS signals reach the

GNSS OVERVIEW

ground, they are very, very weak. We will provide more information about how the user segment deals with this in Chapter 2.

GNSS positioning

> "I have never been lost, but I will admit to being confused for several weeks."
>
> —Daniel Boone, American pioneer and hunter.

If you have a GNSS receiver, it is unlikely that you will ever be lost again. GNSS positioning is based on a process called "trilateration." Simply put, if you don't know your position, but do know your distance from three known points, you can determine your location.

Let's say you are 3 km (1.9 miles) from Person A's house. All you know is that you are on a circle 3 km (1.9 miles) from Person A's house, as shown in **Figure 4**.

But if you also know that you are 4 km (2.5 miles) from Person B's house, you will have a much better idea of where you are since only two places (x and y) exist on both circles, as shown in **Figure 5**.

With a third distance, you can only be in one physical location. If you are 6 km (5.6 miles) from Person C's house, you must be at position X since this is the only location where all three circles (distances) meet.

In Chapter 2, we will show you how the technique of trilateration is extended to GNSS. Conceptually, we are just going to extend the above example by replacing the houses with satellites.

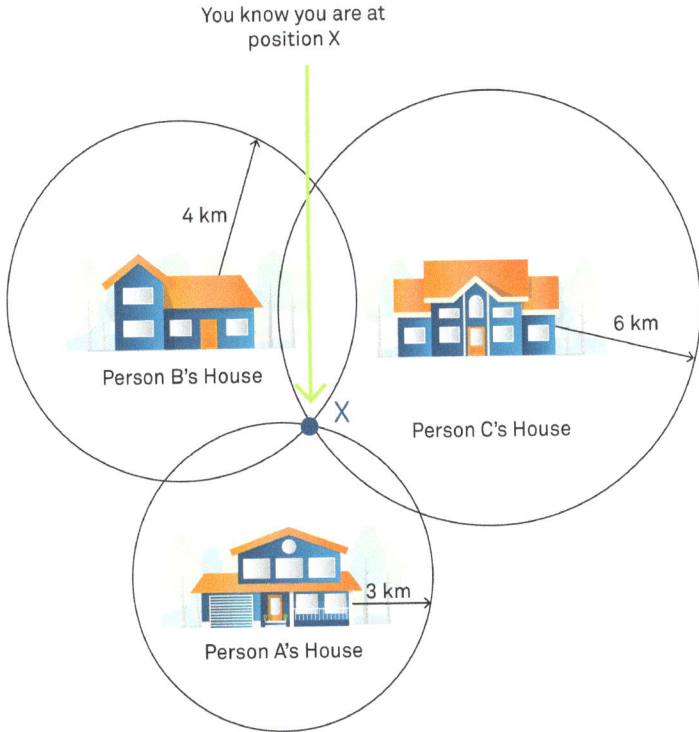

Figure 6 Illustration of trilateration—knowing three distances

An Introduction to GNSS, Third Edition   7

GNSS applications

The first non-military applications of GNSS technology were in surveying and mapping. Today, GNSS is being used for commercial applications in agriculture, transportation, autonomous vehicles, machine control, marine navigation and other industries where efficiencies can be gained from the application of precise, continuously available position and time information. GNSS is also used in a broad range of consumer applications, including vehicle navigation, mobile communications, entertainment and athletics. As GNSS technology improves and becomes less expensive, more and more applications will be conceived and developed.

In addition to position, GNSS receivers can provide users with very accurate time, by "synchronising" their local clock with the high-precision clocks onboard the satellites. This has enabled technologies and applications such as the synchronisation of power grids, cellular systems, the Internet and financial networks.

We'll talk more about GNSS applications in Chapter 9.

GNSS user equipment

The primary components of the GNSS user segment are antennas and receivers, as shown in **Figure 7**. Depending on the application, antennas and receivers may be physically separate or they may be integrated into one assembly.

GNSS antennas

GNSS antennas receive the radio signals that are transmitted by the GNSS satellites and send these signals to the receivers. GNSS antennas are available in a range of shapes, sizes

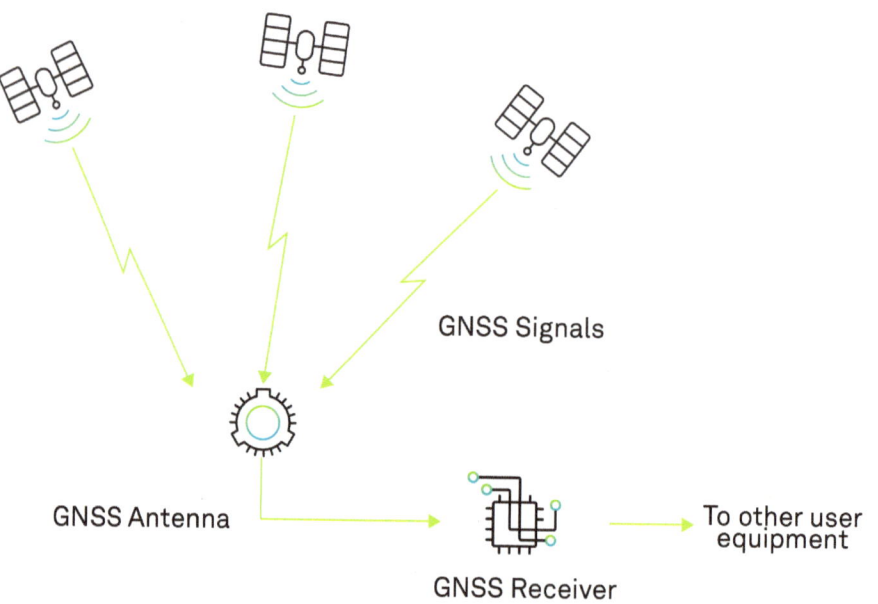

Figure 7 GNSS user equipment

GNSS OVERVIEW

and performances. The antenna is selected based on the application. While a large antenna may be appropriate for a base station, a lightweight, low-profile aerodynamic antenna may be more suitable for aircraft or Unmanned Aerial Vehicles (UAV) installations. **Figure 8** presents a sampling of GNSS antennas.

GNSS receivers

GNSS receivers process the satellite signals recovered by the antenna to calculate position and time. Receivers may be designed to use signals from one or more GNSS constellations. As illustrated in **Figure 9**, receivers are available in many form factors and configurations to meet the requirements of the varied applications of GNSS.

We will talk more about GNSS equipment in Chapter 9.

GNSS augmentation

Positioning based on a standalone GNSS service is accurate to within a few metres (a few yards). However, the accuracy of standalone GNSS may not be adequate for the needs of some users.

Techniques and equipment have been developed to improve the accuracy and availability of GNSS position and time information. We will discuss some of these techniques in Chapter 5.

Closing remarks

Chapter 1 provided an overview of the main concepts and components of GNSS. This high-level summary will help your understanding as we present GNSS in greater detail, starting with a more thorough look at basic GNSS concepts in Chapter 2.

Figure 8 GNSS antennas

Figure 9 GNSS receivers

An Introduction to GNSS, Third Edition   9

"To me, there has never been a higher source of earthly honour or distinction than that connected with advances in science."
—Sir Isaac Newton

Basic GNSS concepts

> "Any sufficiently advanced technology is indistinguishable from magic."
>
> —Arthur C. Clarke, British author, inventor and futurist.

In this chapter, we will introduce basic GNSS concepts. We'll discuss more advanced concepts in subsequent chapters.

GNSS may at first seem like magic, but the more you study and learn about it, the simpler and more elegant it becomes. The basic GNSS concept, shown in **Figure 10**, illustrates the steps involved in using GNSS to determine time and position through to the end-user application.

STEP 1 — SATELLITES:

GNSS satellites orbit the Earth. The satellites know their orbit ephemerides (the parameters that define their orbit) and the time very, very accurately. Ground-based control stations adjust the satellites' ephemerides and time, when necessary.

STEP 2 — PROPAGATION:

GNSS satellites regularly broadcast their ephemerides and time, as well as their status. GNSS radio signals pass through layers of the atmosphere to the user equipment.

STEP 3 — RECEPTION:

GNSS user equipment receives the signals from multiple GNSS satellites and, for each satellite, recovers the information that was transmitted and

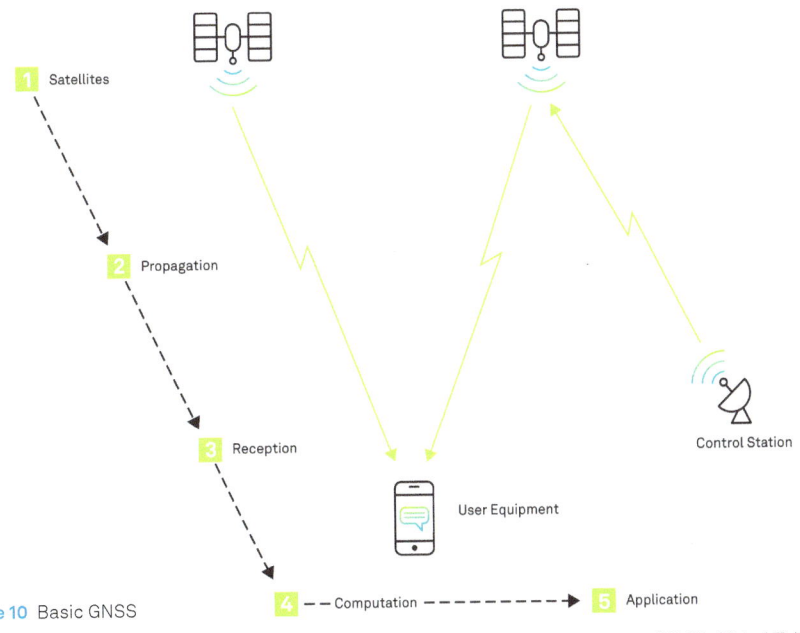

Figure 10 Basic GNSS

BASIC GNSS CONCEPTS

determines the time of propagation (the time it takes the signals to travel from the satellite to the receiver).

STEP 4 — COMPUTATION:

GNSS user equipment uses the recovered information to compute time and position.

STEP 5 — APPLICATION:

GNSS user equipment provides the computed position and time to the end-user application for use in navigation, surveying, mapping and more.

In the following sections, we will discuss each of the above steps in more detail.

Step 1 – Satellites

As outlined in Chapter 1, there are multiple constellations of GNSS satellites orbiting the Earth. A constellation is simply an orderly grouping of satellites in orbits that have been designed to provide a desired coverage, for example, regional or global. We will provide more details about GNSS constellations in Chapter 3.

GNSS satellites orbit well above the atmosphere, about 19,000 to 36,000 kilometres (11,800 to 22,400 miles) above the Earth's surface. They move very fast; GPS satellites move 3.9 km/second.

GNSS satellites are not as small as you might think. One of the current generation of GPS satellites (Block III) weighs over 2,200 kg (4,850 lb), the weight of an average pickup truck. The body of these satellites are 1.8 m x 2.5 m x 3.4 m (5.9' x 8.2' x 11.2') in size. **Figure 11** shows a picture of the body of a Block III GPS satellite to give you a sense of how large they are.

In the relative vacuum of space, satellite trajectories are very stable and predictable. As mentioned, GNSS satellites know their time and orbit ephemerides extremely accurately. If you ask a GPS satellite for the time, it won't tell you 8:30, it will tell you 8:31:39.875921346!

The latest generation of GPS satellites use rubidium clocks that are accurate to within ±5 parts in 10^{11}. These clocks are synchronised by more accurate, ground-based cesium clocks. Cesium clocks are so accurate that they would only gain or lose a second after 100,000 years. By comparison, if you have a quartz watch, it likely has an accuracy of ±5 parts in 10^6 and will lose about a second every two days.

By the way, if all GNSS receivers

Lockheed Martin

Figure 11 Block III GPS satellite

BASIC GNSS CONCEPTS

required a rubidium standard, the viability of GNSS would quickly collapse due to the cost and size of the receiver. Later in the chapter, we will describe the elegant way GNSS systems "transfer" the accuracy of the satellite clocks to GNSS receivers.

You may be wondering why time is such an important factor in GNSS systems. It is because the time it takes a GNSS signal to travel from satellites to receivers is used to determine distances (ranges) to satellites. Accuracy is required because radio waves travel at the speed of light. In one microsecond (a millionth of a second), light travels about 300 metres (328 yards). In a nanosecond (a billionth of a second), light travels about 30 centimetres (1 foot). Small errors in time can result in large errors in calculating position.

GPS was the first GNSS constellation to be launched. The Russian GLONASS, European Galileo and Chinese BeiDou constellations were launched later and are also operational. The benefit to end-users of having access to multiple constellations is accuracy, redundancy and availability. If one system fails, for any reason, GNSS receivers, if they are equipped to do so, can receive and use signals from satellites in other systems. System failure does not happen often, but it's nice to know that if it did, your receiver may still be able to operate.

Regardless, the ability to access multiple constellations is of particular benefit where the line of sight to some of the satellites is obstructed, as is often the case in urban environments or areas with lots of trees.

Satellite orbits

GNSS satellites orbit well above the Earth's atmosphere. GPS and GLONASS satellites orbit at altitudes close to 20,000 km (12,500 miles). BeiDou and Galileo satellites orbit higher, around 21,500 to 36,000 km (13,400 to 22,400 miles) for BeiDou and 23,000 km (14,300 miles) for Galileo. GNSS orbits are more or less circular, highly stable and predictable.

Wind resistance does not affect satellites at 20,000 km (12,500 miles) and higher, but gravitational effects and the pressure of solar radiation do affect GNSS orbits a bit, and the orbits need to be occasionally corrected. While its orbit is being adjusted, a GNSS satellite's status is changed to "out of service" so user equipment knows not to use the affected signals.

Satellite signals

> "Everything should be made as simple as possible, but no simpler."
>
> —Albert Einstein, German-born physicist.

GNSS satellite signals are complex. Describing these signals requires equally complex words like pseudorandom, correlation and Code Division Multiple Access (CDMA). To explain these GNSS concepts, let's first discuss GPS satellite signals.

First and foremost, GPS was designed as a positioning system for the U.S. Department of Defense. To provide high-accuracy position information for military applications, a lot of complexity was designed into the system to make it secure and resistant to jamming and interference (see Chapter 7). Although military and civilian components of GPS are separate, some of the technologies used in the military component have been applied to the civilian component.

Since it achieved full operational capability in December 1993, GPS has

BASIC GNSS CONCEPTS

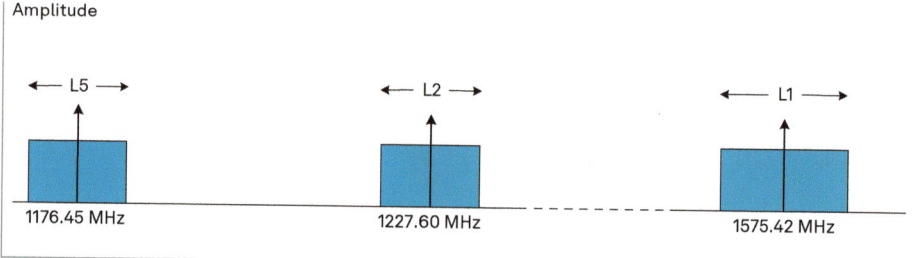

Figure 12 GPS frequency plan

been available to civilian users, who have different requirements for service availability, positioning accuracy and cost. The frequency plans (plans that describe the frequency, amplitude and bandwidth of signals) for each GNSS system are a little different. We will describe these plans in more detail in Chapter 3. To illustrate GNSS concepts, however, we will briefly describe the frequency and signal scheme used by GPS, which is shown in **Figure 12**. Conceptually, this is not much different than the frequency plan for cable or broadcast television channels.

As shown in **Figure 12**, GPS satellites transmit information on the L1, L2 and L5 radio frequencies. You may ask, "How can all GPS satellites transmit on the same frequencies?"

GPS works the way it does because of the transmission scheme it uses, which is called CDMA. CDMA is a form of spread spectrum (meaning the signal is deliberately spread in the frequency domain resulting in a signal with a wider bandwidth). GPS satellite signals, although they are on the same frequency, are modulated by a unique pseudorandom digital sequence or code (PRN code). Pseudorandom means that the signal only appears random; in fact, it actually repeats after a period of time. Receivers know the PRN code for each satellite, because each satellite uses a different code. This allows receivers to correlate (synchronise) with the CDMA signal for a particular satellite. CDMA signals are received at a very low power level, but through this code correlation, the receiver is able to recover the signals and the information they contain.

To illustrate, consider listening to a person in a noise-filled room. Many conversations are taking place, but each conversation is in a different language. You are able to understand the person because you know the language they are speaking. If you are multilingual, you will be able to understand what other people are saying too. CDMA is a lot like this.

You might be interested to learn that Hedy Lamarr, an Austrian-born American scientist and actress, co-invented an early form of spread spectrum communications technology. On August 11, 1942, she and her co-worker, George Antheil, were granted U.S. Patent 2,292,387. Remarkably, Lamarr shifted careers and went on to make 18 films from 1940 to 1949, but the concepts covered in her patent contributed to the development of today's spread spectrum communications.

GPS operates in a frequency band referred to as the L-Band, a portion of

BASIC GNSS CONCEPTS

the radio spectrum between 1 and 2 GHz. L-Band was chosen for several reasons, including:

- Simplification of antenna design. If the frequency had been much higher, user antennas would be more complex.
- Ionospheric delay is more significant at lower frequencies. We'll talk more about ionospheric delay in Step 2 – Propagation, later in this chapter.
- Except in a vacuum, the speed of light is lower at lower frequencies, as evidenced by the separation of the colours in light by a prism. You may have thought the speed of light was a constant at 299,792,458 metres per second (186,000,397 miles per second). It is actually only that exact speed in a vacuum, but through air or any other medium it is slower.
- The coding scheme requires a high bandwidth, which was not available in every frequency band.
- The frequency band was chosen to minimise the effect that weather has on GPS signal propagation.

L1 transmits a navigation message, the coarse acquisition (C/A) code (freely available to the public) and an encrypted precision (P) code, called the P(Y) code (restricted access). The navigation message is a low bit rate message that includes the following information:

- GPS date and time
- Satellite status and health. If the satellite is having problems or its orbit is being adjusted, it will not be usable. When this happens, the satellite will transmit the out-of-service message.
- Satellite ephemeris data, which allows the receiver to calculate the satellite's position. Receivers can determine exactly where the satellite was when it transmitted its time.
- Almanac data, which contains information and status for all GPS satellites, so receivers know which satellites are available for tracking. On start-up, a receiver will recover this "almanac." The almanac consists of coarse orbit and status information for each satellite in the constellation.

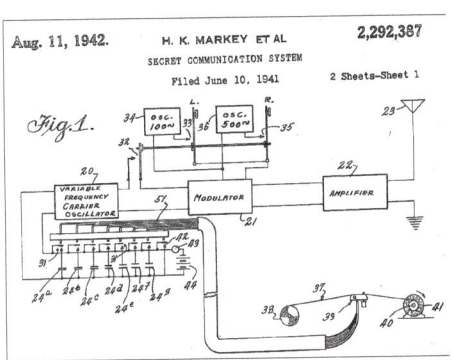

Figure 13 Hedy Lamarr co-authored US Patent 2,292,387 which served as a foundation to the spread spectrum communications we use today

Figure 14 Hedy Lamarr, publicity photo for *Comrade*

An Introduction to GNSS, Third Edition 15

BASIC GNSS CONCEPTS

GPS Date and Time

Satellite Status and Health

Satellite Ephemeris Information

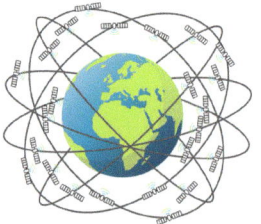
Almanac (information about other satellites in the constellation)

Figure 15 GPS navigation message

The P(Y) code is for military use. It provides better interference rejection than the C/A code, which makes military GPS more robust than civilian GPS. The L2 frequency transmits the P(Y) code and, on newer GPS satellites, it also transmits the C/A code (referred to as L2C), providing a second publicly available code to civilian users. Although the information in the P(Y) code is not accessible to everyone, clever people have figured out ways to use the L2 carrier and code without knowing how it is coded.

While the GPS transmission scheme is complex, it was chosen for many good reasons:

- GPS receivers can recover very weak signals using very small antennas. This keeps the receiver cost low
- Multi-frequency operation allows for ionospheric compensation, since ionospheric delays vary with frequency

- The GPS system is resistant to jamming and interference
- Security: signals used by military applications are not accessible by civilians

Other GNSS systems are conceptually similar to GPS, but there are differences. We will provide more information about these differences in Chapter 3.

Satellite errors

Satellite errors include ephemerides and clock errors. These satellite errors are very, very small, but keep in mind that in one nanosecond, light travels 30 cm (1 foot). These small errors can add up to significant errors in the position calculated by the receiver.

Satellite lifetimes

GNSS satellites don't last forever. Sometimes they are phased out with newer models that have new signals or

BASIC GNSS CONCEPTS

improved timekeeping. Sometimes GNSS satellites do fail and, if they can't be restored, are permanently removed from service.

Satellite corrections

Earth stations continuously monitor the satellites and regularly adjust their time and orbit information to keep the broadcasted information highly accurate. If a satellite's orbit drifts outside the operating limits, it may be taken out of service and its orbit adjusted using small rocket boosters.

In our step-by-step illustration of GNSS in **Figure 10,** the radio signals have left the satellite antenna and are hurtling earthbound at the speed of light.

Step 2 – Propagation

GNSS signals pass through the near-vacuum of space, then through the various layers of the atmosphere to the Earth, as illustrated in **Figure 16**.

To obtain accurate position and time, we need to know the length of the direct path from the satellite to the user equipment (which we refer to as the "range" to the satellite). As shown in **Figure 16**, radio waves do not travel in a straight path. Light travels in a straight line only in a vacuum or through a perfectly homogeneous medium. Just as a straw is seemingly "bent" in a glass of water, radio signals from the satellite are bent as they pass through the Earth's atmosphere. This "bending" increases the amount of time the signal takes to travel from the satellite to the receiver. As we will explain in Step 4, the distance to the satellite is calculated by multiplying the time of propagation (which, you recall, is the time it takes the signals to travel from the satellite to the receiver) by the speed of light. Errors in the propagation time increase or decrease the computed range to the satellite. Incidentally, since the computed range contains errors and is not exactly equal to the actual range, we refer to it as a "pseudorange."

The layer of the atmosphere that most influences the transmission of GPS (and other GNSS) signals is the ionosphere, the layer 50 to 1,000 km (31 to 621 miles) above the Earth's surface. Ultraviolet rays from the sun ionise gas molecules in this layer, releasing free electrons. These electrons influence electromagnetic wave propagation, including GPS satellite signal broadcasts. Ionospheric delays are frequency dependent, so by calculating the range using two frequencies (such as L1 and L2), the receiver can resolve the effect of the ionosphere by comparing the propagation time between the two frequencies.

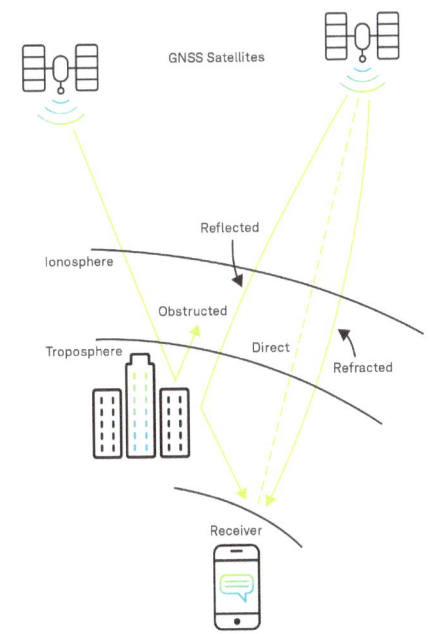

Figure 16 GNSS signal propagation

BASIC GNSS CONCEPTS

The other layer of the atmosphere that influences the transmission of GPS signals is the troposphere, the lowest layer of the Earth's atmosphere. The thickness of the troposphere varies by latitude and season, from about 17 km (10.5 miles) in the middle latitudes up to 20 km (12.4 miles) nearer the equator and thinner at the poles. Tropospheric delay is a function of local temperature, pressure and relative humidity. L1 and L2 are equally delayed by tropospheric conditions, so the effects cannot be eliminated the way ionospheric delay can be. It is possible, however, to model the troposphere to predict and compensate for much of the delay.

Some of the signal energy transmitted by the satellite is reflected on the way to the receiver. This phenomenon is referred to as "multipath propagation." These reflected signals are delayed from the direct signal and, if they are strong enough, can combine with the desired signal, which incorrectly increases the range measurements. Sophisticated techniques have been developed so the receiver tracks the direct signal and ignores multipath signals, which arrive later. In the early days of GPS, most errors came from ionospheric and tropospheric delays, but now more attention is being given to multipath effects, in the interest of continually improving GNSS performance.

Step 3 – Reception

Receivers need at least four satellites to obtain a position (we will explain exactly why in Step 4). The use of more satellites, if they are available, will improve the position solution; however, the receiver's ability to make use of additional satellites may be limited by

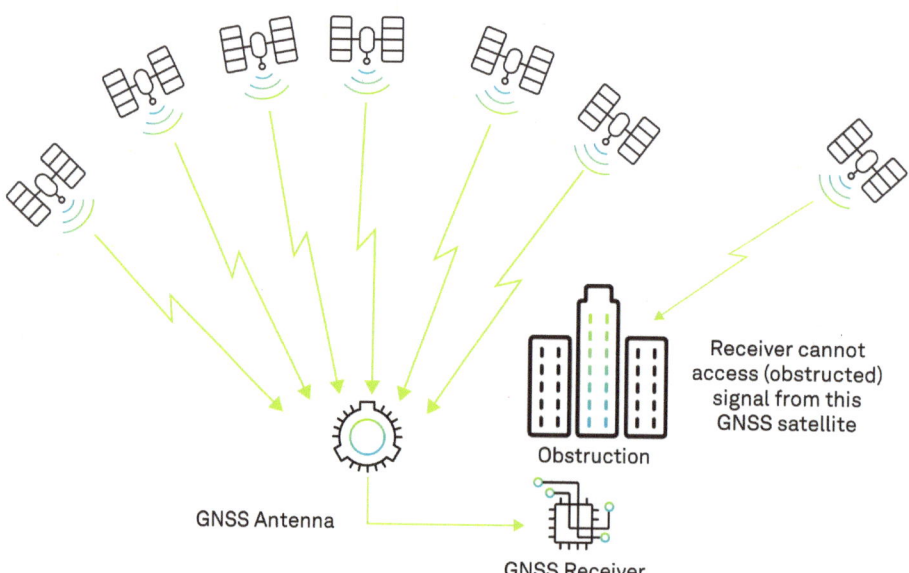

Figure 17 GNSS reception

BASIC GNSS CONCEPTS

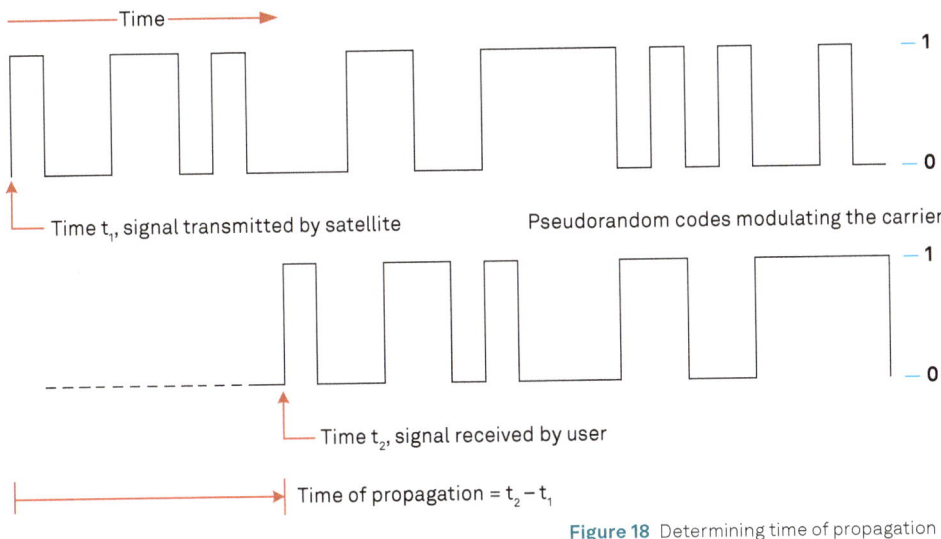

Figure 18 Determining time of propagation

its computational power. The manner by which the receiver uses the additional ranges will generally be the intellectual property of the manufacturer. That said, most user equipment can recover signals from multiple satellites in multiple GNSS constellations.

To determine a position and time, GNSS receivers need to be able to track at least four satellites from one of the GNSS constellations. This means there needs to be a line of sight between the receiver's antenna and the four satellites.

For each satellite being tracked, the receiver determines the propagation time. It can do this because of the pseudorandom nature of the signals. **Figure 18** illustrates the transmission of a pseudorandom code, a series of zeroes and ones. Since the receiver knows the pseudorandom code for each satellite, it can determine the time it received the code from a particular satellite. By comparing the time the signal was received with the transmission time stored in the satellite message, the receiver can determine the time of propagation.

Importance of antenna selection

An antenna behaves both as a spatial and frequency filter; therefore, selecting the right GNSS antenna is critical for optimising performance. An antenna must match the receiver's capabilities and specifications, as well as meet size, weight, environmental and mechanical specifications for the intended application.

Factors to consider when choosing a GNSS antenna include:

1. CONSTELLATION AND SIGNALS

Each GNSS constellation has its own signal frequencies and bandwidths. An antenna must cover the signal frequencies transmitted by the constellation and bandwidth supported by the GNSS receiver.

2. ANTENNA GAIN

Gain is a key performance indicator of a GNSS antenna. Gain can be defined as the relative measure of an antenna's

An Introduction to GNSS, Third Edition

BASIC GNSS CONCEPTS

ability to direct or concentrate radio frequency energy in a particular direction or pattern. A minimum gain is required to achieve a minimum carrier-to-noise ratio (C/No) to track GNSS satellites. The antenna gain is directly related to the overall C/No of the GNSS receivers. Hence, antenna gain helps define the tracking ability of the system.

3. ELEMENT GAIN

The element gain defines how efficient the antenna element is at receiving the signals. In any signal chain, you are only as good as the weakest link, so an antenna element with low element gain might be compensated by an increased low noise amplifier gain. The signal-to-noise ratio or C/No, however, is degraded.

4. ANTENNA BEAMWIDTH & GAIN ROLL-OFF

Gain roll-off is a factor of beamwidth and specifies how much the gain changes over the elevation angle of the antenna. From the antenna's point of view, the satellites rise from the horizon towards zenith (directly overhead) and fall back to the horizon. The variation in gain between zenith and the horizon is known as the gain roll-off. Different antenna technologies have different gain roll-off characteristics. Satellites near the horizon are impacted more by multipath and atmospheric errors.

5. PHASE CENTRE STABILITY

The phase centre of the antenna is the point where the signals transmitted from satellites are collected. When a receiver reports a location fix, that location is essentially the phase centre of the antenna.

The electrical phase centre of any antenna will vary with the position of the transmitting signal it is receiving by as much as a few millimetres (fractions of an inch). As GNSS satellites move across the sky, the electrical phase centre of the signal received will typically move with the satellite position unless the antenna has been carefully designed to minimise Phase Centre Offset (PCO) and Phase Centre Variation (PCV).

The PCO, with respect to the Antenna Reference Point (ARP), is the difference between the mechanical centre of antenna rotation and electrical phase centre location. The PCO is also frequency dependent which means that there can be a different offset for each signal frequency.

The PCV identifies how much the phase centre moves with respect to the satellite elevation angles, see **Figure 19**.

Many users can accept accuracies of less than a metre (one yard), so these small phase centre variations cause a negligible amount of position error. But if you require high precision, such

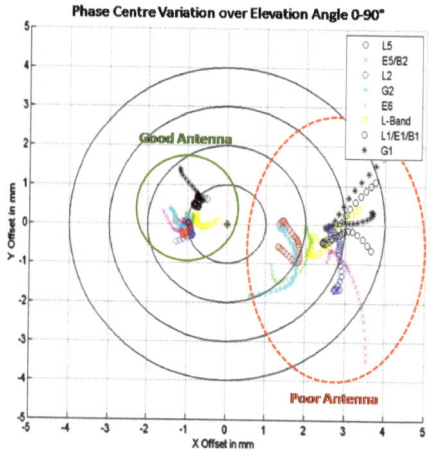

Figure 19 Plot of good and poor antenna phase centre variation over elevation angle

BASIC GNSS CONCEPTS

as real-time kinematic (RTK) receivers that achieve position accuracies of one to four cm (0.4 to 1.6 inches), a few millimetres (fractions of an inch) of phase centre error can translate to a 10-15% error in reported position.

For RTK survey applications, geodetic grade antennas offer superior PCO/PCV performance.

6. APPLICATION

An antenna needs to meet the performance, environmental, mechanical and operational requirements of the intended application. For example, GNSS antennas used for aviation applications should be TSO/FAA certified and be rugged enough to handle extreme temperatures and vibration profiles. Survey rover antennas should be able to survive rough handling by surveyors, including a pole drop.

Step 4 – Computation

If we knew the exact position of three satellites and the exact range to each of them, we would geometrically be able to determine our location. We have suggested that we need ranges to four satellites to determine position. In this section, we will explain why this is and how GNSS positioning actually works.

For each satellite being tracked, the receiver calculates how long the satellite signal took to reach it, as follows:

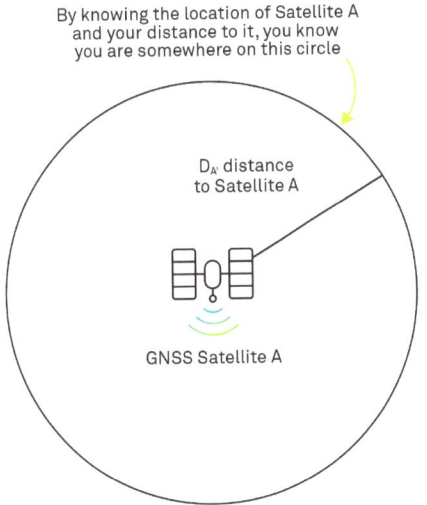

Figure 20 Ranging to first satellite

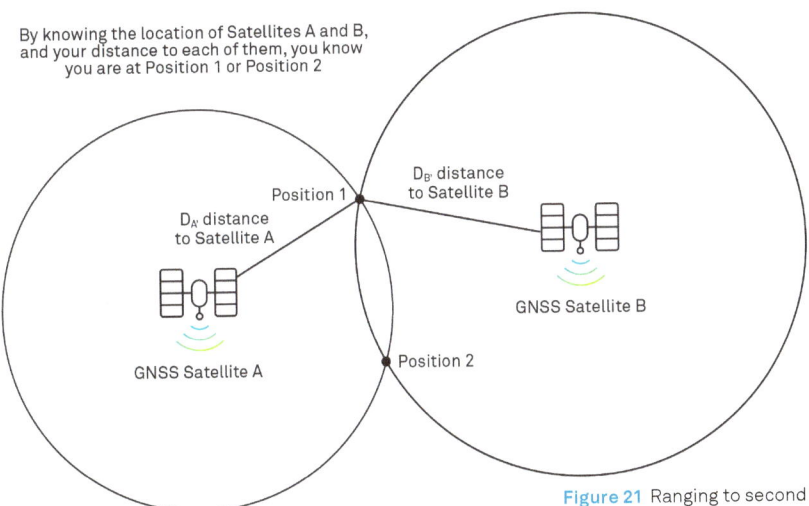

Figure 21 Ranging to second satellite

An Introduction to GNSS, Third Edition

BASIC GNSS CONCEPTS

**Propagation Time =
Time Signal Reached Receiver −
Time Signal Left Satellite**

Multiplying this propagation time by the speed of light gives the distance to the satellite.

For each satellite being tracked, the receiver now knows where the satellite was at the time of transmission (because the satellite broadcasts its orbit ephemerides) and it has determined the distance to the satellite when it was there. Using trilateration, a method of geometrically determining the position of an object in a manner similar to triangulation, the receiver calculates its position.

To help us understand trilateration, we'll present the technique in two dimensions. The receiver calculates its range to Satellite A. As we mentioned, it does this by determining the amount of time it took for the signal from Satellite A to arrive at the receiver and multiplying this time by the speed of light. Satellite A communicated its location (determined from the satellite orbit ephemerides and time) to the receiver, so the receiver knows it is somewhere on a circle with a radius equal to the range and centred at the location of Satellite A, as illustrated in **Figure 20**. In three dimensions, we would show ranges as spheres, not circles.

The receiver also determines its range to a second satellite, Satellite B. Now the receiver knows it is at the intersection of two circles, at either Position 1 or 2, as shown in **Figure 21**.

You may be tempted to conclude that ranging to a third satellite would be required to resolve your location to Position 1 or Position 2. However, one of the positions can most often be eliminated as not feasible because, for example, it is in space or in the middle of the Earth. You might also be tempted to extend our illustration to three dimensions and suggest that only three ranges are needed for positioning but

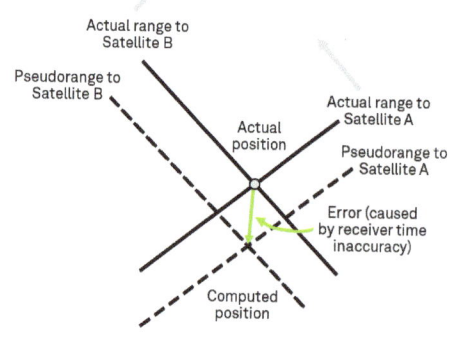

Figure 22 Position error

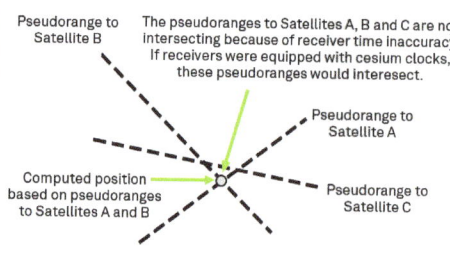

Figure 23 Detecting position error

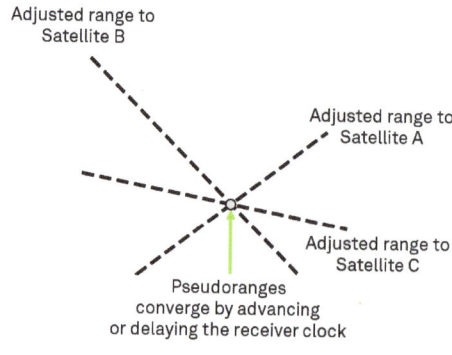

Figure 24 Convergence

BASIC GNSS CONCEPTS

four ranges are actually necessary. Why is this?

It turns out that receiver clocks are not nearly as accurate as the clocks onboard the satellites. Most are based on quartz crystals. Remember, we said these clocks were accurate to only about five parts per million. If we multiply this by the speed of light, it results in an accuracy of ±1,500 metres (1,640 yards). When we determine the range to two satellites, our computed position, will be out by an amount proportional to the inaccuracy in our receiver clock, as illustrated in **Figure 22**.

We want to determine our actual position but, as shown in **Figure 22**, the receiver time inaccuracy causes range errors that result in position errors. The receiver knows there is an error; it just doesn't know the size of the error. If we now compute the range to a third satellite, it doesn't intersect the computed position, as shown in **Figure 23**.

Now for one of the ingenious techniques used in GNSS positioning.

The receiver knows the reason the pseudoranges to the three satellites are not intersecting is because its clock is not very good. The receiver is programmed to advance or delay its clock until the pseudoranges to the three satellites converge at a single point, as shown in **Figure 24**.

The incredible accuracy of the satellite clock has been "transferred" to the receiver clock, greatly reducing the receiver clock error in the position determination. The receiver now has both an accurate position and extremely accurate time. This presents opportunities for a broad range of applications, as we'll discuss.

The above technique shows how, in a two-dimensional representation, receiver time inaccuracy can be almost eliminated and position determined using ranges to three satellites. When we extend this technique to three dimensions, we need to add a range to a fourth satellite. This is the reason why line-of-sight to a minimum of four GNSS satellites is needed to determine position.

GNSS error sources

A GNSS receiver calculates position based on data received from satellites. However, there are many sources of errors that, if left uncorrected, cause the position calculation to be inaccurate.

The type of error and how it is mitigated is essential to calculating precise position, as the level of precision is only useful to the extent that the measurement can be trusted. This book dedicates three chapters to this important topic. Chapter 4 presents key sources of GNSS errors while Chapter 5 discusses methods of error resolution and impact on accuracy and other performance factors. Chapter 9 presents the equipment and network infrastructure necessary to generate and receive correction data.

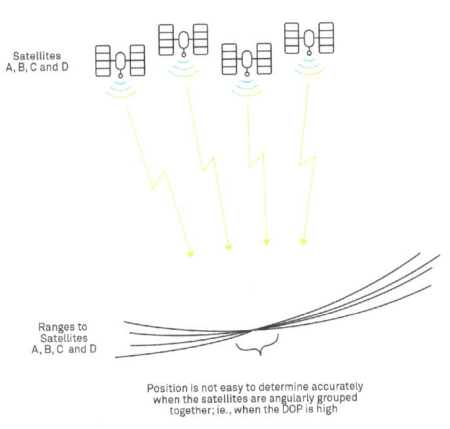

Figure 25 Dilution of precision (poor satellite geometry)

An Introduction to GNSS, Third Edition   23

BASIC GNSS CONCEPTS

Dilution of precision (DOP)

The geometric arrangement of satellites, as they are presented to the receiver, affects the accuracy of position and time calculations. Receivers are ideally designed to use signals from all available satellites in a manner that minimises this so-called "dilution of precision."

To illustrate DOP, consider the example shown in **Figure 25**, where the satellites being tracked are clustered in a small region of the sky. As you can see, it is difficult to determine where the ranges intersect. The position is "spread" over the area of range intersections, which is enlarged by range inaccuracies (this can be viewed as a "thickening" of the range lines).

As shown in **Figure 26**, the addition of a range measurement to a satellite that is angularly separated from the cluster allows the receiver to determine position more precisely.

Although DOP is calculated using complex statistical methods, we can say the following: DOP is a numerical representation of satellite geometry and is dependent on the locations of satellites visible to the receiver.

The smaller the value of DOP, the more precise the time or position calculation. The relationship is shown in the following formula:

Inaccuracy of Computed Position = DOP x Inaccuracy of Range Measurement

So, if DOP is very high, the inaccuracy of the computed position will be much larger than the inaccuracy of the range measurement.

- A DOP above six results in generally unacceptable accuracies for Differential (D) GNSS and RTK operations

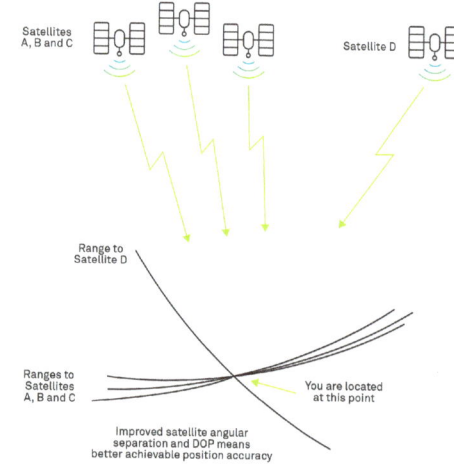

Figure 26 Dilution of precision (improved geometry)

- DOP varies with time of day and geographic location
- DOP can be calculated without determining the range. All that is needed is the satellite positions and the approximate receiver location
- Adding more satellites causes the DOP to decrease

DOP can be expressed as a number of separate elements that define the DOP for a particular type of measurement, for example, HDOP (Horizontal Dilution of Precision), VDOP (Vertical Dilution of Precision), and PDOP (Position Dilution of Precision). These factors are mathematically related. In some cases, for example, when satellites are low in the sky, HDOP is low, and it will therefore be possible to get a good to excellent determination of horizontal position (latitude and longitude), but VDOP may only be adequate for a moderate altitude determination.

In Canada and in other countries at high

latitudes, GNSS satellites are lower in the sky, so achieving optimal DOP for some applications, particularly where good VDOP is required, can be a challenge.

Applications where the available satellites are low on the horizon or angularly clustered, such as those in urban environments or in deep open-pit mining, may expose users to the pitfalls of DOP.

Step 5 – Application

Once the errors have been accounted for in the GNSS equation, the receiver can determine its position and time and pass this information on to the end-user application. The GNSS technology market is a ubiquitous, multi-billion-dollar industry. Applications range from simple hand-held metre-level (yard-level) navigation aids to robust, centimetre-level (inch-level) positioning solutions for survey, military and autonomous applications. With users demanding GNSS positioning functionality in increasingly challenging environments, GNSS technology is being integrated with other sensors such as inertial technology to enhance positioning capabilities and dependability. We look at a variety of sensors in Chapter 6.

As applications become more complex and ubiquitous, the likelihood of "GNSS denied" scenarios, such as jamming and spoofing, increases, whether intentional or unintentional. Chapter 7 discusses the causes and mitigation techniques of GNSS denial. Our final chapter, Chapter 9, promises to inspire with some of our most exciting customer applications.

Closing remarks

This has been a tough chapter, and we're pleased you persevered through the basics of GNSS positioning. Chapter 3 provides additional information about the GNSS constellations that have been implemented or are planned. Chapters 4, 5, 6, 7 and 8 discuss advanced GNSS concepts, and Chapter 9 discusses equipment and applications — how the simple outputs of this incredible technology are being used.

"Look deep into nature, and then you will understand everything better."
—Albert Einstein

3 GNSS constellations

"The dinosaurs became extinct because they didn't have a space program."
—**Larry Niven,** American science fiction author.

Figure 27 Launch of Galileo satellite

Larry Niven is suggesting that if the dinosaurs had had a space program, they could have intercepted and deflected the asteroid that may have led to their extinction.

Unlike the dinosaurs, several countries now have existing or planned space programs that include the implementation of national or regional GNSS. In this chapter, we will provide an overview of these systems.

GNSS CONSTELLATIONS

Figure 28 GPS IIR-M satellite (artist's rendition)

Satellites	31
Orbital planes	6
Orbit inclination	55 degrees
Orbit radius	20,200 km (12,552 miles)

Table 1 GPS satellite constellation

GPS (Global Positioning System), United States

GPS was the first GNSS system. GPS (or NAVSTAR, as it is officially called) satellites were first launched in the late 1970s and early 1980s for the U.S. Department of Defense. Since that time, several generations (referred to as "Blocks") of GPS satellites have been launched. Initially, GPS was available only for military use but in 1983, a decision was made to extend GPS to civilian use. A GPS satellite is depicted in **Figure 28**.

Space segment

The GPS space segment is summarised in **Table 1**. The orbit period of each satellite is approximately 12 hours, so this provides a GPS receiver with at least six satellites in view from most points on Earth, under open-sky conditions, and typically more.

A GPS satellite orbit is illustrated in **Figure 29**.

GPS satellites continually broadcast their signals, which contain satellite ephemeris data, ranging signals, identification, clock data and almanac data. The satellites are identified either by their Space Vehicle Number (SVN) or their Pseudorandom Noise (PRN) code.

Signals

Table 2 provides further information on GPS signals. GPS signals are based on CDMA (Code Division Multiple Access) technologies, which we discussed in Chapter 2.

GNSS CONSTELLATIONS

Designation	Frequency	Description
L1	1575.42 MHz	L1 is modulated by the C/A code (coarse/acquisition) which is available to all users and the P-code (precision) which is encrypted for military and other authorised users. Beginning with the Block III satellites, L1 is also modulated with the L1C (civilian) code and is discussed later in this chapter.
L2	1227.60 MHz	L2 is modulated by the P-code and, beginning with the Block IIR-M satellites, the L2C (civilian) code. L2C broadcasts civil navigation (CNAV) messages and is discussed later in this chapter under *GPS Modernisation*.
L5	1176.45 MHz	L5, available beginning with Block IIF satellites, broadcasts CNAV messages. The L5 signal is discussed later in this chapter under *GPS Modernisation*.

Table 2 GPS signal characteristics

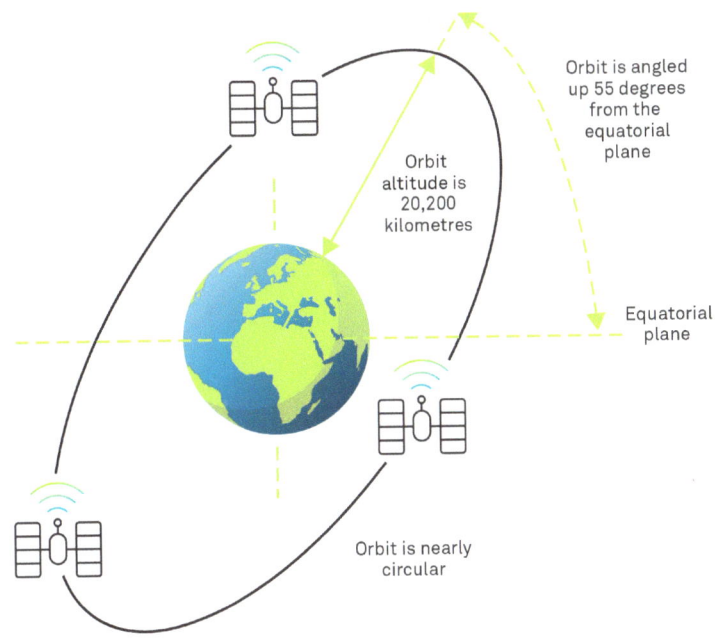

Figure 29 GPS satellite orbit

An Introduction to GNSS, Third Edition   29

GNSS CONSTELLATIONS

Figure 30 GPS control segment

Master control station	Schriever AFB
Alternate master control station	Vandenberg AFB
Air Force monitor stations	Schriever AFB, Cape Canaveral, Hawaii, Ascension Island, Diego Garcia, Kwajalein
AFSCN remote tracking stations	Schriever AFB, Vandenberg AFB, Hawaii, New Hampshire, Greenland, United Kingdom, Diego Garcia, Guam
NGA monitor stations	USNO Washington, Alaska, United Kingdom, Ecuador, Uruguay, South Africa, Bahrain, South Korea, Australia, New Zealand
Ground antennas	Cape Canaveral, Ascension Island, Diego Garcia, Kwajalein

Control segment

The GPS control segment consists of a master control station (and a backup master control station), monitor stations, ground antennas and remote tracking stations, as shown in **Figure 30**.

There are 16 monitor stations located throughout the world, six from the U.S. Air Force and ten from the NGA (National Geospatial-Intelligence Agency, also part of the U.S. Department of Defense). The monitor stations track the satellites via their broadcast signals, which contain satellite ephemeris data, ranging signals, identification, clock data and almanac data. These signals are passed to the master control station, where the ephemerides are recalculated. The resulting ephemerides and timing corrections are transmitted back up to the satellites through data uploading stations.

GNSS CONSTELLATIONS

The ground antennas are co-located with monitor stations and used by the Master Control Station to communicate with and control the GPS satellites.

The Air Force Satellite Control Network (AFSCN) remote tracking stations provide the Master Control Station with additional satellite information to improve telemetry, tracking and control.

GPS modernisation

GPS reached Fully Operational Capability (FOC) in 1995. In 2000, a project was initiated to modernise the GPS space and ground segments to take advantage of new technologies and user requirements.

Space segment modernisation includes new signals, as well as improvements in atomic clock accuracy, satellite signal strength and reliability. Control segment modernisation includes improved ionospheric and tropospheric modelling and in-orbit accuracy and additional monitoring stations. User equipment has also evolved to take advantage of space and control segment improvements.

L2C

The modernised GPS satellites (Block IIR-M and later) are transmitting a new civilian signal, designated L2C, ensuring the accessibility of two civilian codes. L2C is easier for the user segment to track than L2 P(Y) and it delivers improved navigation accuracy. It also provides the ability to directly measure and remove the ionospheric delay error for a particular satellite by using the civilian signals on both L1 and L2. The L2C signal is available from 24 satellites in 2022 and is expected to be available on all satellites by 2026.

L5

The United States has implemented a third civilian GPS frequency (L5) at 1176.45 MHz. The modernised GPS satellites (Block IIF and later) are transmitting L5.

The benefits of the L5 signal include meeting the requirements for critical safety-of-life applications such as those needed for civil aviation and providing:

- Improved ionospheric correction
- Signal redundancy
- Improved signal accuracy
- Improved interference rejection

The L5 signal is available from 16 satellites in 2021 and is expected to be available on all satellites by 2030.

L1C

A fourth civilian GPS signal, L1C, is available from the next generation Block III GPS satellites. The European Galileo, Japanese QZSS, Indian NavIC and Chinese BeiDou systems already broadcast or plan to broadcast L1C compatible signals, making it a future standard for international interoperability.

L1C features a new modulation scheme that will improve GPS reception in cities and other challenging environments. L1C is available from five satellites as of 2022 and is expected to be available on all satellites by 2034.

Other

In addition to the new L1C, L2C and L5 signals, GPS satellite modernisation includes new military signals.

GNSS CONSTELLATIONS

GLONASS (Global Navigation Satellite System), Russia

GLONASS was developed by the Soviet Union as an experimental military communications system during the 1970s. When the Cold War ended, the Soviet Union recognised that GLONASS had commercial applications through the system's ability to transmit weather broadcasts, communications, navigation and reconnaissance data.

The first GLONASS satellite was launched in 1982, and the system was declared fully operational in 1993. After a period where GLONASS performance declined, Russia committed to bringing the system up to the required minimum of 18 active satellites. Currently, GLONASS has a full deployment of 24 satellites in the constellation.

GLONASS satellites have evolved since the first ones were launched. The latest generation, GLONASS-M, is shown in **Figure 31**, being readied for launch.

GLONASS space segment

The GLONASS space segment is summarised in **Table 3** and consists of 24 satellites, in three orbital planes, with eight satellites per plane.

The GLONASS constellation geometry repeats about once every eight days. The orbit period of each satellite is approximately 8/17 of a sidereal[1] day so that, after eight sidereal days, the GLONASS satellites have completed exactly 17 orbital revolutions.

Each orbital plane contains eight equally spaced satellites. One of the

Figure 31 GLONASS-M satellite in final manufacturing

Satellites	24 plus 3 spares
Orbital planes	3
Orbital inclination	64.8 degrees
Orbit radius	19,140 km (11,893 miles)

Table 3 GLONASS satellite constellation

---

[1] A sidereal day is the time it takes for one complete rotation of the Earth, relative to a particular star. A sidereal day is about four minutes shorter than a mean solar day.

GNSS CONSTELLATIONS

Designation	Frequency	Description
L1	1598.0625–1609.3125 MHz	L1 is modulated by the HP (High Precision) and the SP (Standard Precision) signals.
L2	1242.9375–1251.6875 MHz	L2 is modulated by the HP and SP signals. The SP code is identical to that transmitted on L1.
L3OC	1202.025 MHz	L3 is a civilian signal based on CDMA.

Table 4 GLONASS signal characteristics

satellites will be at the same spot in the sky at the same sidereal time each day.

The satellites are placed into nominally circular orbits with target inclinations of 64.8 degrees and an orbital radius of 19,140 kilometres (11,893 miles), about 1,060 km (659 miles) lower than GPS satellites.

The GLONASS satellite signal identifies the satellite and includes:
- Positioning, velocity and acceleration information for computing satellite locations
- Satellite health information
- Offset of GLONASS time from UTC (SU) (Universal Time Coordinated of Russia)
- Almanac of all GLONASS satellites

GLONASS control segment

The GLONASS control segment consists of the system control centre and a network of command tracking stations across Russia. Similar to that of GPS, the GLONASS control segment monitors the health of the satellites, determines the ephemeris corrections, as well as the satellite clock offsets with respect to GLONASS time and UTC. Twice a day, it uploads corrections to the satellites.

Figure 32 View of Earth (as seen by Apollo 17

> "The Earth was absolutely round... I never knew what the word 'round' meant until I saw Earth from space."
>
> – **Alexei Leonov**, Soviet astronaut, talking about his historic 1985 spacewalk.

GNSS CONSTELLATIONS

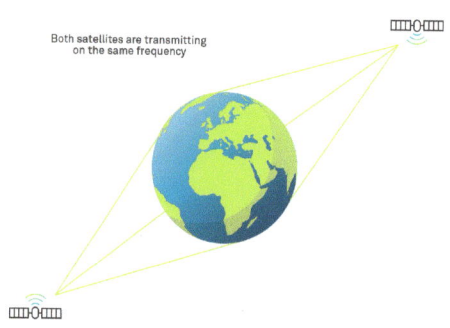

Figure 33 GLONASS antipodal satellites

GLONASS signals

Table 4 summarises the GLONASS signals.

Each GLONASS satellite transmits on a slightly different L1 and L2 frequency, with the P-code (HP code) and the C/A code (SP code) on both L1 and L2. GLONASS satellites transmit the same code at different frequencies, a technique known as FDMA (frequency division multiple access). Note that this is a different technique from that used by GPS and the other GNSS constellations.

GLONASS signals have the same polarisation (orientation of the electromagnetic waves) as GPS signals and have comparable signal strength.

The GLONASS system is based on 24 satellites using 14 frequencies. The satellites can share the frequencies by having antipodal satellites transmitting on the same frequency. Antipodal satellites are in the same orbital plane but are separated by 180 degrees. The paired satellites can transmit on the same frequency because they will never appear at the same time in view of a receiver on the Earth's surface, as shown in **Figure 33**.

GLONASS modernisation

As the current GLONASS-M satellites reach the end of their service life, they will be replaced with next-generation GLONASS-K satellites. The new satellites will provide the GLONASS system with new GNSS signals.

L3OC

The first block of GLONASS-K satellites (GLONASS-K1) broadcast the new civilian signal, designated L3, centred at 1202.025 MHz. Unlike the existing GLONASS signals, L3OC is based on CDMA, which will ease interoperability with other GNSS constellations.

L3OC is available from 47 satellites as of 2022.

L1OC and L2OC

The second block of GLONASS-K satellites (GLONASS-K2) adds two more CDMA-based signals broadcast at the L1OC and L2OC frequencies. The exiting FDMA L1 and L2 signals will continue to be broadcast as well to support legacy receivers.

L5OCM

The third block of GLONASS-K satellites (GLONASS-KM) will add an L5 CDMA signal to the GLONASS system, as well as new L1 and L3 signals. These will be at the same frequencies used by other GNSS constellations and named L1OCM, L3OCM and L5OCM.

BeiDou Navigation Satellite System (China)

The Chinese GNSS is known as the BeiDou Navigation Satellite System (BDS).

Phase 2 of the BeiDou system (BDS-2) officially became operational in December 2012, providing coverage

GNSS CONSTELLATIONS

for the Asia Pacific region. Initially, the regional BeiDou space segment had five Geostationary Earth Orbit (GEO)[2] satellites, five Inclined Geosynchronous Orbit (IGSO) satellites and four Medium Earth Orbit (MEO) satellites.

Phase 3 of the BeiDou system (BDS-3) became fully operational in 2020, and BeiDou now provides global coverage with enhanced regional coverage. As of 2021, the space segment consists of a constellation of 45 satellites, as shown in **Table 5** and **Table 6**.

BeiDou signals

The BeiDou signals, based on CDMA technology, are summarised in **Table 7** and **Table 8**.

Figure 34 In Chinese, the Big Dipper Constellation is known as BeiDou

Galileo (European Union)

In May 1999, a mountaineering expedition carried a GPS receiver to the summit of Mount Everest, allowing them to accurately measure its elevation at

Satellites	2 GEO	7 IGSO	3 MEO
Orbital inclination	–	55 degrees	55 degrees
Orbit radius	35,787 km (22,237 miles)	35,787 km (22,237 miles)	21,528 km (13,377 miles)

Table 5 BDS-2 satellites in constellation

Satellites	2 GEO	4 IGSO	26 MEO
Orbital planes	–	–	3
Orbital inclination	–	55 degrees	55 degrees
Orbit radius	35,787 km (22,237 miles)	35,787 km (22,237 miles)	21,528 km (13,377 miles)

Table 6 BDS-3 satellites in constellation

---

2 A geosynchronous orbit has an orbital period matching the Earth's sidereal rotation period. This synchronisation means that for an observer at a fixed location on Earth, a satellite in a geosynchronous orbit returns to exactly the same place in the sky at exactly the same time each day. The term geostationary is used to refer to the special case of a geosynchronous orbit that is circular (or nearly circular) and at zero (or nearly zero) incli-

GNSS CONSTELLATIONS

Designation	Frequency	Description
B1I	1561.098 MHz	B1I provides both public service signals and restricted service signals.
B2I	1207.140 MHz	B2I provides both public service signals and restricted service signals.
B3I	1268.520 MHz	B3I provides both public service signals and restricted service signals.

Table 7 BDS-2 satellite signal characteristics

Designation	Frequency	Description
B1I	1561.098 MHz	B1I provides both public service signals and restricted service signals.
B1C	1575.42 MHz	B1C provides public service signals. GEO satellites will broadcast BDSBAS corrections on B1C.
B2a	1176.45 MHz	B2A provides public service signals. GEO satellites will broadcast BDSBAS corrections on B2a.
B2b	1207.140 MHz	B2b provides short message communication services, international search and rescue service and PPP service.
B3I	1268.520 MHz	B3I provides both public service signals and restricted service signals.

Table 8 BDS-3 satellite signal characteristics

GNSS CONSTELLATIONS

Figure 35 Galileo Galilei

> "Measure what is measurable, and make measurable what is not so."
>
> –**Galileo Galilei**, Italian physicist, mathematician, astronomer and philosopher.

8,850 metres (29,035 ft). We think Galileo, the Italian physicist, would have been happy.

Galileo, Europe's Global Navigation Satellite System, provides a highly accurate and guaranteed global positioning service under civilian control. The U.S. and European Union have been cooperating since 2004 to ensure that GPS and Galileo are compatible and interoperable at the user level.

Galileo guarantees availability of service under all but the most extreme circumstances, and it will inform users, within seconds, of a failure of any satellite. This makes it suitable for applications where safety is crucial, such as in air and ground transportation.

The first experimental Galileo satellite, part of the Galileo System Test Bed (GSTB) was launched in December 2005, and a second in April 2008. The purpose of these experimental satellites was to characterise critical Galileo technologies, which were already in development under European Space Agency (ESA) contracts. Four in-orbit validation satellites were launched in 2011 and 2012 to validate the basic Galileo space and ground segment.

System design

The Galileo space segment is summarised in **Table 9**. Galileo navigation

Figure 36 Galileo satellite in orbit

Satellites	24 operational
Orbital planes	3
Orbital inclination	56 degrees
Orbit radius	23,222 km (14,430 miles)

Table 9 Galileo satellite constellation

GNSS CONSTELLATIONS

signals provide coverage at all latitudes. The large number of satellites, together with the optimisation of the constellation and the availability of the three active spare satellites, ensures that the loss of one satellite has no discernible effect on the user segment.

Two Galileo Control Centres (GCC), located in Europe, control the Galileo satellites. Data recovered by a global network of thirty Galileo Sensor Stations (GSS) is sent to the GCC through a redundant communications network. The GCCs use the data from the sensor stations to compute integrity information and to synchronise satellite time with ground station clocks. Control centres communicate with the satellites through uplink stations, which are installed around the world.

Galileo provides a global Search and Rescue (SAR) function based on the operational search and rescue satellite-aided Cospas-Sarsat[3] system. To do this, each Galileo satellite is equipped with a transponder that transfers distress signals to the Rescue Coordination Centre

---

[3] Cospas-Sarsat is an international satellite-based Search And Rescue (SAR) distress alert detection and information distribution system, established by Canada, France, United States and the former Soviet Union in 1979.

Designation	Frequency	Description
E1 A	1575.42 MHz	Public regulated service signal.
E1 B		Safety-of-life and open service signal (data).
E1 C		Safety-of-life and open service signal (dataless).
E5a I	1176.45 MHz	Open service signal (data).
E5a Q		Open service signal (dataless).
E5b I	1207.14 MHz	Safety-of-life and open service signal (data).
E5b Q		Safety-of-life and open service signal (dataless).
AltBOC	1191.795 MHz	Combined E5a/E5b signal.
E6 A	1278.75 MHz	Public regulated service signal.
E6 B		High accuracy service signal (data).
E6 C		Commercial service signal (dataless).

Table 10 Galileo signal characteristics

Service	Description
Free Open Service (OS)	Provides positioning, navigation and precise timing service. It is available for use by any person with a Galileo receiver. No authorisation is required to access this service.
Highly Reliable Commercial Service (CS)	Service providers can provide added-value services, for which they can charge the end customer. The CS signal provides high data throughput and accurate authenticated data relating to these additional commercial services.
Safety-of-Life Service (SOL)	Improves on the Open Service by providing timely warnings to users when it fails to meet certain margins of accuracy. A service guarantee is provided for this service.
Government Encrypted Public Regulated Service (PRS)	Highly encrypted restricted-access service offered to government agencies that require a high availability navigation signal.
High Accuracy Service (HAS)	HAS is a PPP service with orbit corrections, clock corrections, between-signal biases (code, phase) and atmospheric corrections. It would include corrections for all Galileo signals and GPS L1, L2C and L5. The target is 20 cm (8 inches) 95% horizontal user error within 300 seconds, everywhere and for free. It is broadcast on E6. This service is planned for SIS testing in 2021, initial service in 2022 and full service in 2024.
Open Service Navigation Message Authentication (OSNMA)	Galileo will provide OSNMA service to allow users to authenticate GNSS navigation data. This service has been tested since 2021 and is expected to be in service in 2023.

Table 11 Galileo services

GNSS CONSTELLATIONS

Designation	Frequency	Description
L5	1176.45 MHz	L5 will be modulated with the SPS and RS signals.
S	2492.028 MHz	S will be modulated with the SPS and RS signals. Navigation signals will also be transmitted on S.

Table 12 NavIC signal characteristics

(RCC), which then initiates the rescue operation. At the same time, the system provides a signal to the user, informing them that their situation has been detected and that help is underway. This latter feature is new and is considered a major upgrade over existing systems, which do not provide feedback to the user.

Galileo signals

Table 10 provides further information about Galileo signals.

Galileo services

Five Galileo services are available, as summarised in **Table 11**.

NavIC (Navigation with Indian Constellation), India

The NavIC system (formerly known as IRNSS) provides coverage for India and the surrounding region. The system provides positioning accuracy of better than 10 metres (10.9 yards) throughout India and better than 20 m (21.9 yards) for the area surrounding India by 1,500 km (932 miles).

NavIC provides two services. A Standard Positioning Service (SPS) is available to all users, and a Restricted Service (RS) is available to authorised users only.

Table 12 summarises the NavIC signals. Plans to broadcast an additional signal, L1, from all NavIC satellites was announced by the Indian government in 2021.

The first NavIC satellite was launched in July of 2013. As of 2022, the NavIC system consists of eight satellites, three of them in geostationary orbits and five in inclined geosynchronous orbits, with plans to expand to 12 satellites.

QZSS (Quasi-Zenith Satellite System), Japan

QZSS is a four-satellite system that provides regional communication services and positioning information for the mobile environment. The focus of this system is for the Japan region, but it will provide service to the Asia-Oceania region.

QZSS provides limited accuracy in standalone mode, so it is viewed as a GPS augmentation service. The QZSS satellites use the same frequencies as GPS and have clocks that are synchronised with GPS time. This allows the QZSS satellites to be used as if they were additional GPS satellites. QZSS satellites also broadcast a signal compatible with Satellite-Based

GNSS CONSTELLATIONS

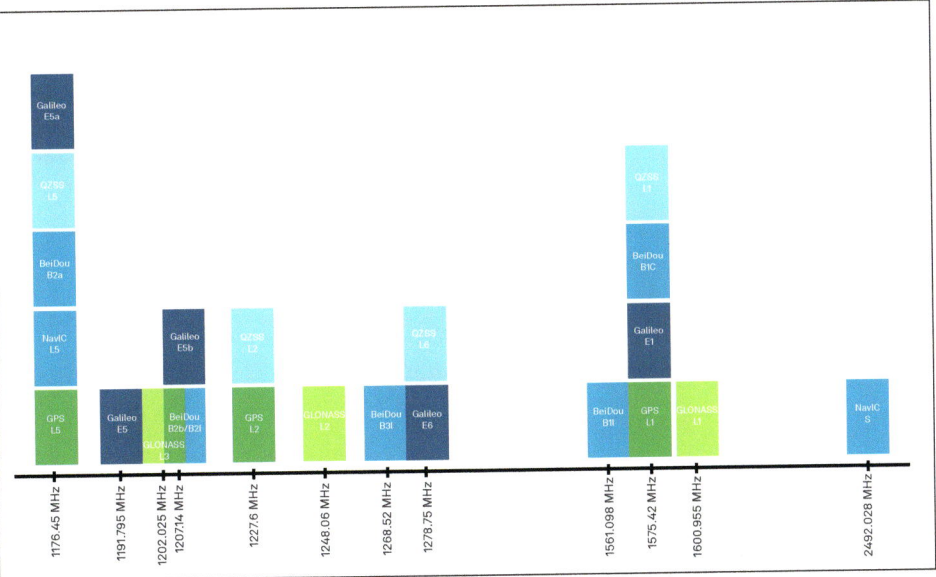

Figure 37 GNSS signals

Augmentation Systems (SBAS) and a high-precision signal at L6. See Chapter 5 to learn more about SBAS.

Three of the QZSS satellites are in a periodic Quasi-Zenith Orbit (QSO). These orbits will allow the satellites to "dwell" over Japan for more than 12 hours a day, at an elevation above 70° (meaning they appear almost overhead most of the time).

Japan intends to expand the QZSS system to a seven-satellite system by 2024.

GNSS signal summary

As more GNSS constellations and signals become available, the more complex the GNSS spectrum becomes. **Figure 37** shows the signals for the current GNSS systems.

Closing remarks

Now that you know more about Global Navigation Satellite Systems, we will discuss advanced GNSS concepts in the following chapters.

"The diversity of the phenomena of nature is so great, and the treasures hidden in the heavens so rich, precisely in order that the human mind shall never be lacking in fresh nourishment."

–Johannes Kepler

GNSS error sources

In Chapter 2, we introduced the concept of GNSS error sources. These are the factors that make it difficult for a GNSS receiver to calculate an exact position. In this chapter, we will look more deeply into these error sources.

Contributing source	Error range	
Satellite clocks	±2 m	(6.6 feet)
Orbit errors	±2.5 m	(8.2 feet)
Ionospheric delays	±5 m	(16.4 feet)
Tropospheric delays	±0.5 m	(1.6 feet)
Receiver noise	±0.3 m	(1 foot)
Multipath	±1 m	(3.3 feet)

Table 13 GNSS sources of errors

Satellite clocks

The atomic clocks in the GNSS satellites are very accurate, but they do drift a small amount. Unfortunately, a small inaccuracy in the satellite clock results in a significant error in the position calculated by the receiver. For example, 10 nanoseconds of clock error results in 3 metres (3.3 yards) of position error.

The clock on the satellite is monitored by and compared to the even more accurate clock used in the GNSS ground control system. In the downlink data, the satellite provides the user with an estimate of its clock offset. Typically, the estimate has an accuracy of about ±2 metres (±2.2 yards), although this can vary between different GNSS systems. To obtain a more accurate position, the GNSS receiver needs to compensate for the clock error.

One way of compensating for clock error is to download precise satellite clock information from a Space Based Augmentation System (SBAS) or Precise Point Positioning (PPP) service provider. The precise satellite clock information contains corrections for the clock errors that are calculated by the SBAS or PPP system. More information about SBAS and PPP is provided in Chapter 5.

Another way of compensating for clock error is to use a Differential GNSS (DGNSS) or Real-Time Kinematic (RTK) receiver configuration. Chapter 5 discusses DGNSS and RTK in depth.

Orbit errors

GNSS satellites travel in very precise, well-known orbits. However, like the satellite clock, the orbits do vary a small amount. Also, like the satellite clocks, a small variation in the orbit results in a significant error in the position calculated.

The GNSS ground control system continually monitors satellite orbits. When a satellite orbit changes, the ground control system sends a correction to the satellites, and the satellite ephemeris is updated. Even with the corrections from the GNSS ground control system, there are still small errors in the orbit that can result in up to ±2.5 metres (2.7 yards) of

GNSS ERROR SOURCES

position error.

Similar to clock errors, one way of compensating for satellite orbit errors is to download precise ephemeris information from an SBAS or PPP service provider.

Another way of compensating for satellite orbit errors is to use a DGNSS or RTK receiver configuration. More information about SBAS, PPP, DGNSS and RTK is provided in Chapter 5.

Ionospheric delay

The ionosphere is the layer of atmosphere between 50 and 1,000 km (31 to 621 miles) above the Earth. This layer contains electrically charged particles called ions. These ions alter the transmission time of the satellite signals and can cause a significant amount of satellite position error (typically ±5 metres (5.5 yards), but can be more during periods of high ionospheric activity).

Ionospheric delay varies with solar activity, time of year, season, time of day and location. This makes it very difficult to predict how much ionospheric delay is impacting the calculated position.

Ionospheric delay also varies based on the radio frequency of the signal passing through the ionosphere. GNSS receivers that can receive more than one GNSS signal, L1 and L2 for example, can use this to their advantage. By comparing the measurements for L1 to the measurements for L2, the receiver can determine the amount of ionospheric delay and remove this error from the calculated position.

For receivers that can only track a single GNSS frequency, ionospheric models are used to reduce ionospheric delay errors. Due to the varying nature of ionospheric delay, models are not as effective as using multiple frequencies at removing ionospheric delay.

Ionospheric conditions are very similar within a local area, so the base station and rover receivers experience a very similar delay. This allows DGNSS and RTK systems to compensate for ionospheric delay.

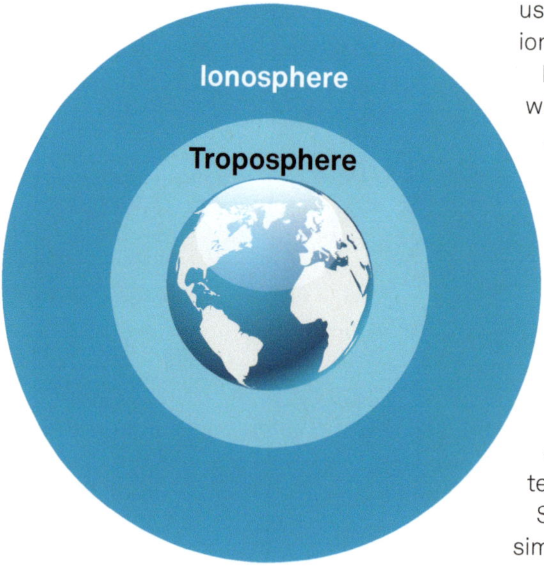

Tropospheric delay

The troposphere is the layer of atmosphere closest to the surface of the Earth.

Variations in tropospheric delay are caused by changing humidity, temperature and atmospheric pressure.

Since tropospheric conditions are very similar within a local area, base station and rover receivers experience a very similar delay. This allows DGNSS and RTK systems to compensate for tropospheric delay.

Figure 38 Ionosphere and troposphere

GNSS receivers can also use tropospheric models to estimate the amount of error caused by tropospheric delay.

Receiver noise

Receiver noise refers to the position error caused by the GNSS receiver hardware and software. High-end GNSS receivers tend to have less receiver noise than lower-cost GNSS receivers.

Multipath

Multipath error occurs when a signal from the same satellite reaches a GNSS antenna via two or more paths, such as reflected off the wall of a building. Because the reflected signal travels further to reach the antenna, the reflected signal arrives at the receiver slightly delayed. This effect is seen in **Figure 39**. This delayed signal can cause the receiver to calculate an incorrect position.

The simplest way to reduce multipath errors is to place the GNSS antenna in a location that is away from the reflective surface. When this is not possible, the GNSS receiver and antenna must manage the multipath signals.

Long delay multipath errors are typically handled by the GNSS receiver, while short delay multipath errors are handled by the GNSS antenna. Due to the additional technology required to deal with multipath signals, high-end GNSS receivers and antennas tend to be better at rejecting multipath errors.

Closing remarks

This chapter has described the error sources that cause inaccuracies in the calculation of position. In Chapter 5, we will describe the methods that GNSS receivers use to mitigate these errors and provide a more accurate position.

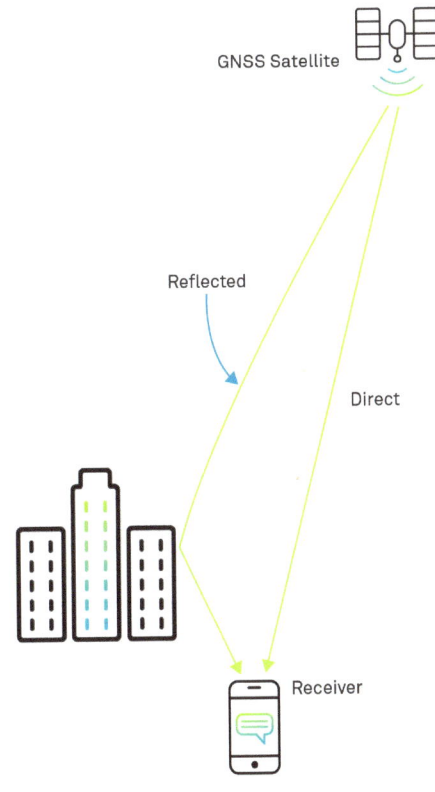

Figure 39 Multipath illustration

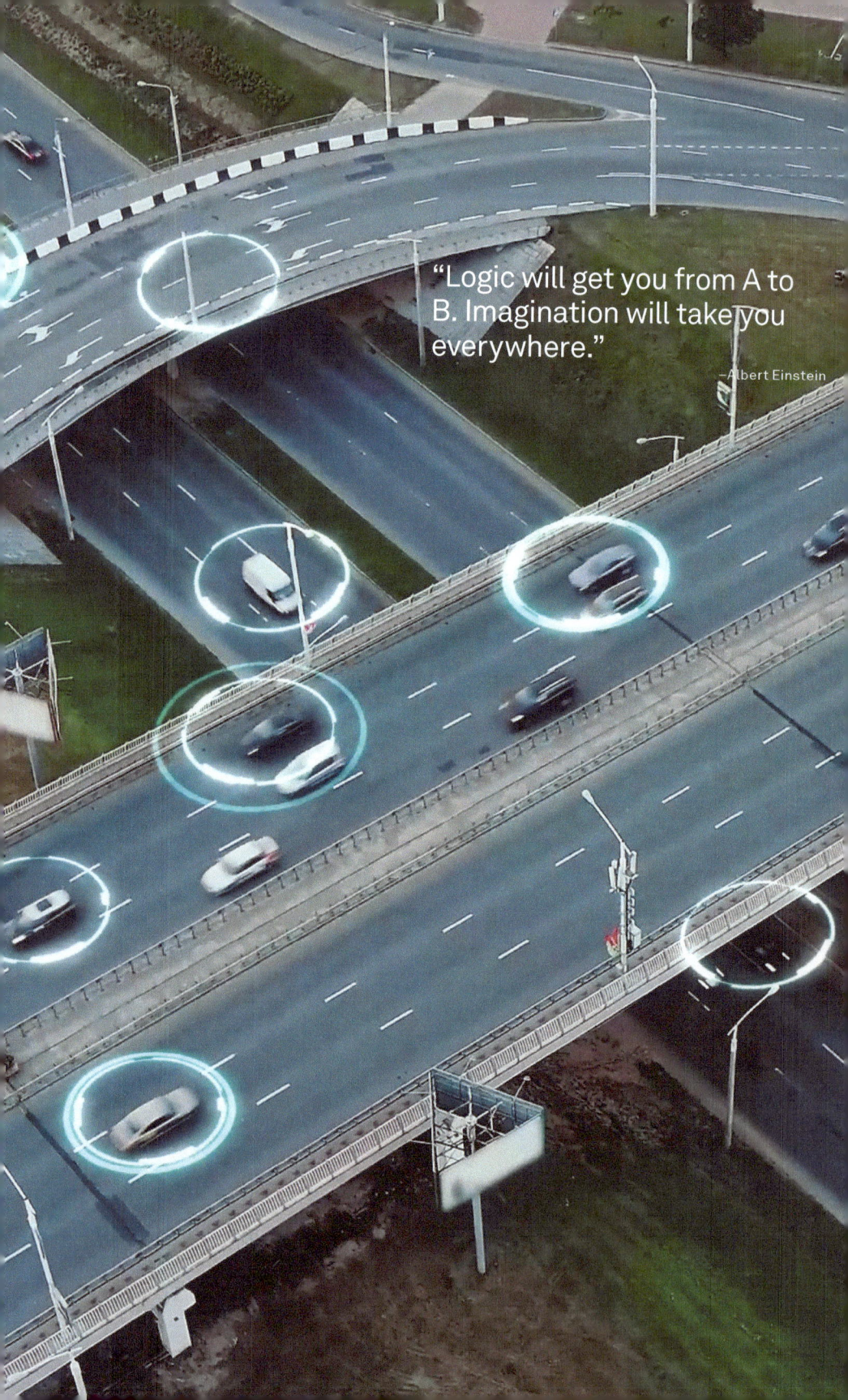

Resolving errors

Resolving errors is fundamental to the performance of a GNSS receiver. How a manufacturer develops a receiver, including both hardware and software design elements, directly impacts the effectiveness of error resolution. The more errors a receiver can eliminate, the higher the degree of positioning accuracy and reliability it can achieve.

What is the ideal technique to correct for errors? There really is no "best way," as it all depends on the positioning performance required by the end-user application. Using the GNSS receiver in your smartphone to find that new restaurant does not require the same level of performance as landing an unmanned helicopter on a moving platform, for example.

There are trade-offs between the different methods of removing errors in GNSS signals. The methods employed depend on the unique requirements of each application, such as level of accuracy, system complexity, solution availability, reliability and cost.

In Chapter 2, we introduced the basic concepts of GNSS positioning, specifically as they apply to single-point positioning, where a single GNSS receiver operates individually, or "standalone," to determine its location and time. In this chapter, we introduce methods by which GNSS receivers improve performance by using more advanced techniques that mitigate or eliminate errors within the position calculation. Fundamentally GNSS positioning all starts with the simple mathematical formula of: Velocity = Distance ÷ Time. Therefore, factors that affect the distance to the satellite or the time it takes for a satellite signal to arrive at the antenna need to be addressed.

Luckily, some very smart people have developed techniques to resolve errors. In general, these techniques can be described as follows:

A. Modelling of the phenomena that are causing the errors and estimating the correction values.
B. Reducing or removing the error sources by differencing between receivers.

In this chapter, we will examine a number of correction techniques, how they work and some of the benefits and challenges of each method. But let's first look at the concepts of multi-frequency/multi-constellation and code versus carrier phase GNSS measurements and their impact on error resolution and positioning performance.

Multi-frequency, multi-constellation

The ability of a GNSS receiver to handle multiple frequencies from multiple constellations in the calculation of position is essential to optimal error resolution.

Multi-frequency

Using multi-frequency receivers is the most effective way to remove ionospheric error from the position calculation. Ionospheric error varies with frequency, so it impacts the various GNSS signals differently. By comparing the delays of two GNSS signals, L1 and L2 for example, the receiver can correct for the impact of ionospheric errors.

The modernised wideband signals

RESOLVING ERRORS

provide inherent noise and multipath mitigation capabilities. When receivers combine modern wideband signal capabilities with the ability to remove ionospheric error using dual-frequency, significant improvements in both measurement and positioning accuracy can be achieved.

Multi-frequency receivers also provide more immunity to interference. If there is interference in the L2 frequency band around 1227.60 MHz, a multi-frequency receiver will still track L1 and L5 signals to ensure ongoing positioning.

Multi-constellation

As described previously, a multi-constellation receiver can access signals from several constellations: GPS, GLONASS, BeiDou and Galileo, for example. The use of other constellations in addition to GPS results in a larger number of satellites in the field of view, which has the following benefits:

- Reduced signal acquisition time
- Improved position and time accuracy
- Reduced problems caused by obstructions such as buildings and foliage
- Improved spatial distribution of visible satellites, resulting in improved dilution of precision

When a receiver utilises signals from a variety of constellations, protection from signal blockage is built into the solution. If a signal is blocked due to the surrounding environment, there is a very high likelihood that the receiver can simply pick up a signal from another constellation — ensuring solution continuity. While extremely rare, if a GNSS system fails, there are other systems available.

To determine a position in GNSS-only

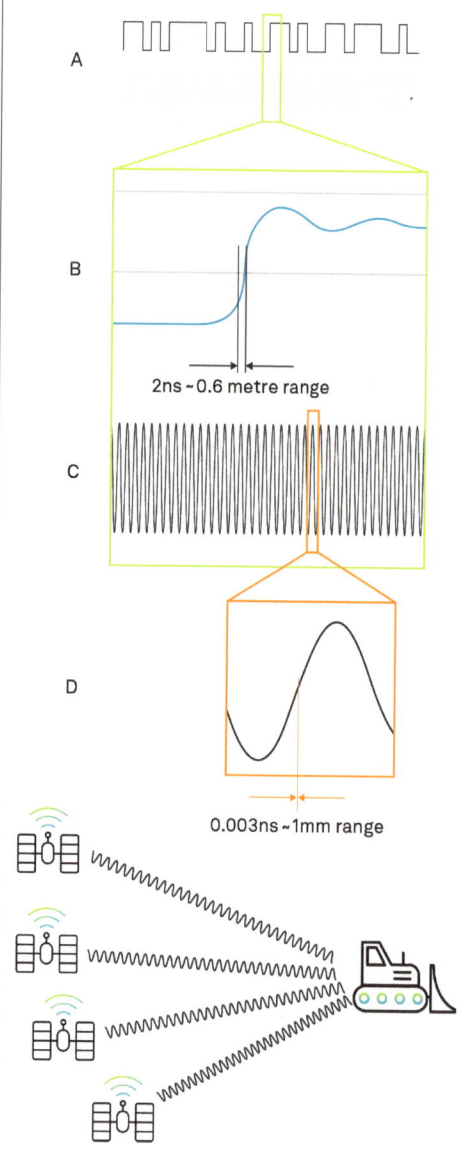

Figure 40 Pseudorange vs carrier phase

mode, a receiver must track a minimum of four satellites. In multi-constellation mode, the receiver is more likely to find enough satellites to track, even in challenging GNSS environments.

GNSS measurements: pseudorange and carrier phase

The positioning technique described in Chapter 2 is referred to as a pseudorange-based technique because the receiver uses the ranges derived from the correlation of the satellite PRN codes. This results in positioning accuracies of a few metres or yards. For some applications, such as surveying, higher accuracies are required. Carrier-based techniques such as Real-Time Kinematic (RTK) and Precise Point Positioning (PPP) have been developed to provide positions that are orders of magnitude more accurate than pseudorange-based GNSS.

A. Phase modulation of the carrier wave using the PRN code is used to differentiate satellite signals and to provide signal timing information for range measurements. See item A in **Figure 40**.

B. Measurements based on the PRN modulation are unambiguous, but precision is limited to sub-metre or foot level. See item B in **Figure 40**.

C. The carrier wave for the GNSS signal is a sine wave with a period of less than one metre (one yard) (19 cm [7.5 inches] for L1), allowing for more precise measurements. See item C in **Figure 40**.

D. Measurements of the carrier wave's phase can be made to millimetre (sub-inch) precision, but the measurement is ambiguous because the total number of cycles between the satellite and

receiver is unknown. See item D in **Figure 40**.

Resolving or estimating the carrier phase ambiguities is key to achieving precise positioning with RTK or PPP. The two methods use different techniques to achieve this, but both make use of:

- Pseudorange (code-based) position estimates
- Mitigation of positioning errors, either by using relative positioning or correction data
- Multiple satellite signal observations to find the ambiguity terms that fit best with the measurement data

Therefore, the method employed by the receiver, code-only or code-and-carrier-based measurements, impacts the positioning performance.

Differential GNSS (DGNSS)

A commonly used technique for improving GNSS performance is DGNSS, which is illustrated in **Figure 41**.

In DGNSS, the position of a fixed GNSS receiver, referred to as a base station, is determined to a high degree of accuracy using conventional surveying techniques. Then, the base station determines ranges to the GNSS satellites in view using:

- The code-based positioning technique described in Chapter 2
- The location of the satellites determined from the precisely known orbit ephemerides and satellite time

The base station compares the surveyed position to the position calculated from the satellite ranges. Differences between the positions can be attributed to satellite ephemeris and clock errors, but mostly to errors associated with atmospheric delay. The base station uses the difference between

RESOLVING ERRORS

the calculated and surveyed position to generate a correction message which it sends to other receivers (rovers). The rover receivers incorporate the corrections from the base station into their position calculations to improve position accuracy.

If corrections need to be applied in real-time, differential positioning requires both a data link between the base station and rovers and at least four GNSS satellites in view at both the base station and the rovers. The absolute accuracy of the rover's computed position will depend on the absolute accuracy of the base station's position.

Since GNSS satellites orbit high above the Earth (approx. 20,200 km or 12,550 miles), the propagation paths from the satellites to the base stations and rovers pass through similar atmospheric conditions, as long as the base station and rovers are not too far apart. DGNSS works very well with maximum separations of up to tens of kilometres or miles.

Satellite Based Augmentation System (SBAS)

For applications where the cost of a differential GNSS system is not justified, or if the rover stations are spread over too large an area, an SBAS may be more appropriate for enhancing position accuracy.

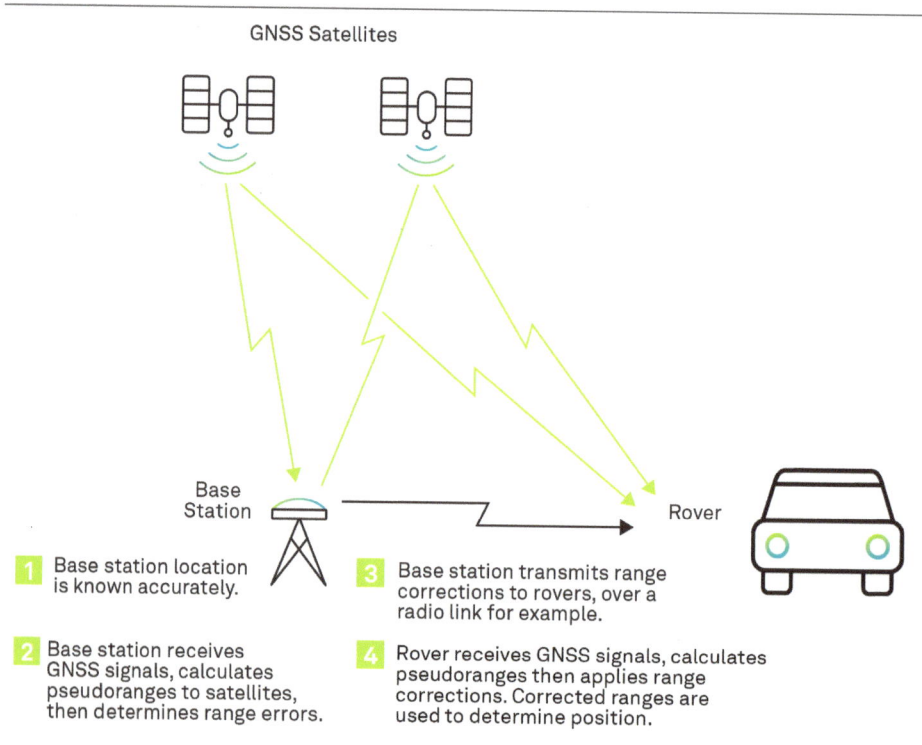

Figure 41 Differential GNSS (DGNSS) overview

RESOLVING ERRORS

SBAS are geosynchronous satellite systems that provide services for improving the accuracy, integrity and availability of basic GNSS signals:

- Accuracy is enhanced through the transmission of wide-area corrections for GNSS range errors
- Integrity is enhanced by the SBAS network quickly detecting satellite signal errors and sending alerts to receivers that they should not track the failed satellite
- Signal availability can be improved if the SBAS transmits ranging signals from its satellites

SBAS include reference stations, master stations, uplink stations and geosynchronous satellites, as shown in **Figure 42**.

Reference stations, which are geographically distributed throughout the SBAS service area, receive GNSS signals and forward them to the master station. Since the locations of the reference stations are accurately known, the master station can accurately calculate wide-area corrections.

Corrections are uplinked to the SBAS satellite and then broadcast to GNSS receivers throughout the SBAS coverage area.

User equipment receives the corrections and applies them to range calculations.

The following sections provide an overview of some of the SBAS services that have been implemented around the world or that are planned.

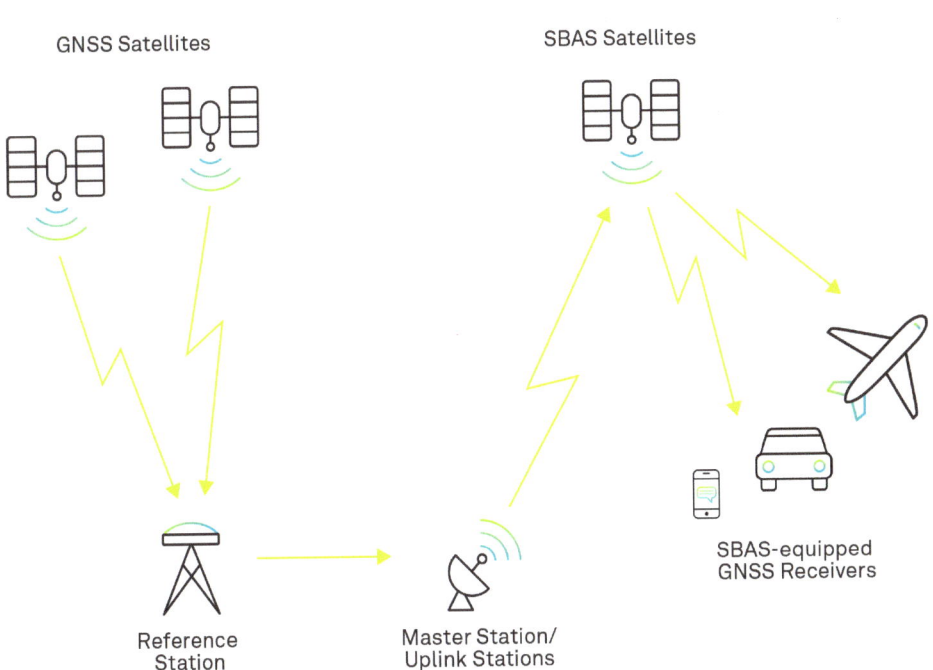

Figure 42 Satellite-Based Augmentation System (SBAS) overview

RESOLVING ERRORS

Wide Area Augmentation System (WAAS)

The U.S. Federal Aviation Administration (FAA) has developed the WAAS to provide GPS corrections and a certified level of integrity for the aviation industry, enabling aircraft to conduct precision approaches to airports. The corrections are also available free of charge to civilian users in North America.

A Wide-Area Master Station (WMS) receives GPS data from Wide-Area Reference Stations (WRS) located throughout North America. The WMS calculates differential corrections and then uplinks these to two WAAS geostationary satellites for broadcast across North America.

Separate corrections are calculated for ionospheric delay, satellite timing and satellite orbits, which allows error corrections to be processed separately, if appropriate, by the user application.

The WAAS broadcasts correction data on the same frequency as GPS, which allows for the use of the same receiver and antenna equipment as that used for GPS. To receive correction data, user equipment must have line of sight to one of the WAAS satellites.

European Geostationary Navigation Overlay Service (EGNOS)

The European Space Agency, in cooperation with the European Commission (EC) and EUROCONTROL (European organisation supporting European aviation), has developed the EGNOS, an augmentation system that improves the accuracy of positions derived from GPS signals and alerts users about the reliability of the GPS signals.

The EGNOS system augments GPS signals over Europe and North Africa. EGNOS transmits an open service to the EU member states, plus Norway and Sweden, and a safety-of-life service to the European Civil Aviation Conference (ECAC) Flight Information Regions.

In a future upgrade, the EGNOS system will also support Galileo signals.

BeiDou Satellite-Based Augmentation System (BDSBAS)

The BDSBAS system is an extension of the BeiDou GNSS system. It provides accuracy improvements and integrity service to users in China and the surrounding area.

MTSAT Satellite Based Augmentation System (MSAS)

MSAS is an SBAS that provides augmentation services to Japan. It uses two Multi-functional Transport Satellites (MTSAT) and a network of ground stations to augment GPS signals in Japan.

GPS-Aided GEO Augmented Navigation (GAGAN) system

GAGAN is an SBAS that supports flight navigation over Indian airspace. The system is based on three geostationary satellites, 15 reference stations installed throughout India, three uplink stations and two control centres. GAGAN is compatible with other SBAS systems, such as WAAS, EGNOS and MSAS.

System for Differential Corrections and Monitoring (SDCM)

The Russian Federation is developing SDCM to provide Russia with accuracy improvements and integrity monitoring for both the GLONASS and GPS navigation systems. The Russian Federation plans

RESOLVING ERRORS

to also provide Precise Point Positioning (PPP) services for L1/L3 GLONASS.

Other SBASs

The Korean Augmentation Satellite System (KASS) is an SBAS being developed by South Korea.

The Solución de Aumentación para Caribe, Centro y Sudamérica (SACCSA) is planned for the South America, Central America and Caribbean region.

The A-SBAS is intended to provide service for Africa and Indian Ocean.

The Southern Positioning Augmentation System (SouthPAN) is an SBAS being developed for Australia and New Zealand.

Ground Based Augmentation System (GBAS)

A GBAS provides differential corrections and satellite integrity monitoring to receivers using a very high frequency (VHF) radio link. Also known as a Local Area Augmentation System (LAAS), a GBAS consists of several GNSS antennas placed at known locations, a central control system and a VHF radio transmitter.

GBAS covers a relatively small area (by GNSS standards) and is used for applications that require high levels of accuracy, availability and integrity. Airports are an example of a GBAS application.

Real-Time Kinematic (RTK)

The positioning technique we described in Chapter 2 is referred to as code-based positioning because the receiver correlates with and uses the pseudorandom codes transmitted by four or more satellites to determine

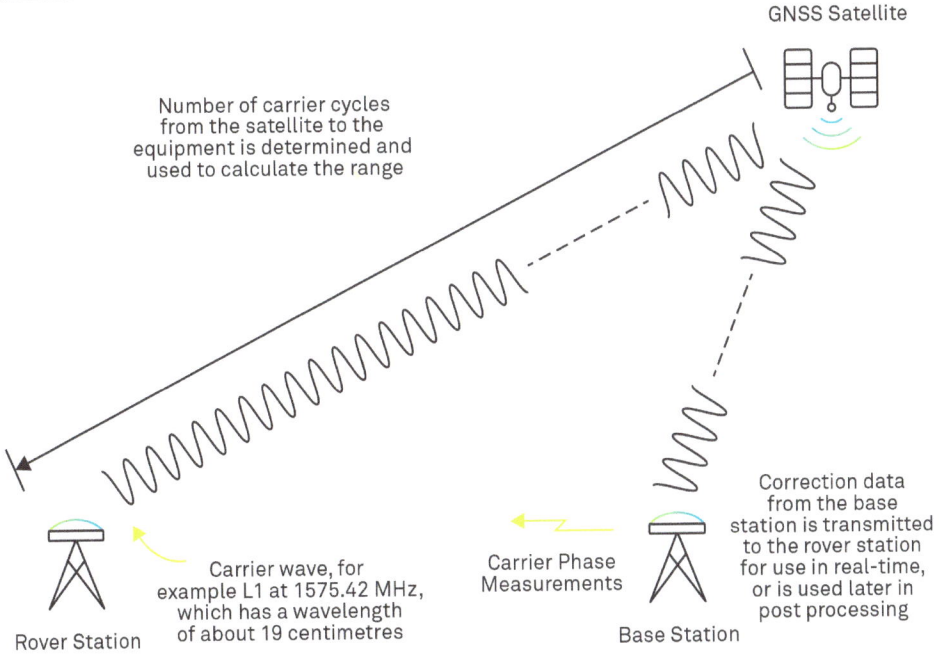

Figure 43 Real-Time Kinematic (RTK) overview

RESOLVING ERRORS

the ranges to the satellites. From these ranges and knowing where the satellites are, the receiver can establish its position to within a few metres or yards.

For applications that require higher accuracies, RTK is a technique that uses carrier-based ranging and provides ranges (and therefore positions) that are orders of magnitude more precise than those available through code-based positioning.

RTK techniques are complicated. The basic concept is to reduce and remove errors common to a base station and rover pair, as illustrated in **Figure 43**.

At a very basic conceptual level, the range is calculated by determining the number of carrier cycles between the satellite and the rover station, then multiplying this number by the carrier wavelength.

The calculated ranges still include errors from such sources as satellite clock and ephemerides, and ionospheric and tropospheric delays. To eliminate these errors and to take advantage of the precision of carrier-based measurements, RTK performance requires measurements to be transmitted from the base station to the rover station.

A complicated process called "ambiguity resolution" is needed to determine the number of whole cycles. Despite being a complex process, high-precision GNSS receivers can resolve the ambiguities almost instantaneously. For a brief description of ambiguities, see the *GNSS measurements–code and carrier phase precision* section earlier in this chapter.

Rovers determine their position using algorithms that incorporate ambiguity resolution and differential correction. Like DGNSS, the position accuracy achievable by the rover depends on, among other things, its distance from the base station (referred to as the "baseline") and the accuracy of the differential corrections. Corrections are as accurate as the known location of the base station and the quality of the base station's satellite observations. Site selection is important for minimising environmental effects such as interference and multipath, as is the quality of the base station and rover receivers and antennas.

Network RTK

Network RTK is based on the use of several widely spaced permanent stations. Depending on the implementation, positioning data from the permanent stations is regularly communicated to a central processing station. RTK user terminals transmit their approximate location to the central station, which calculates and transmits correction information or corrected position to the RTK user terminal on-demand. The benefit of this approach is an overall reduction in the number of RTK base stations required. Depending on the implementation, data may be transmitted over cellular radio links or other wireless mediums.

Precise Point Positioning (PPP)

PPP is a positioning technique that removes or models GNSS errors to provide a high level of position accuracy from a single receiver. A PPP solution depends on a set of corrections generated from a network of global reference stations. Once the corrections are calculated, they are delivered to the end-user via satellite or over the Internet. These corrections are used by the receiver to improve position accuracy.

RESOLVING ERRORS

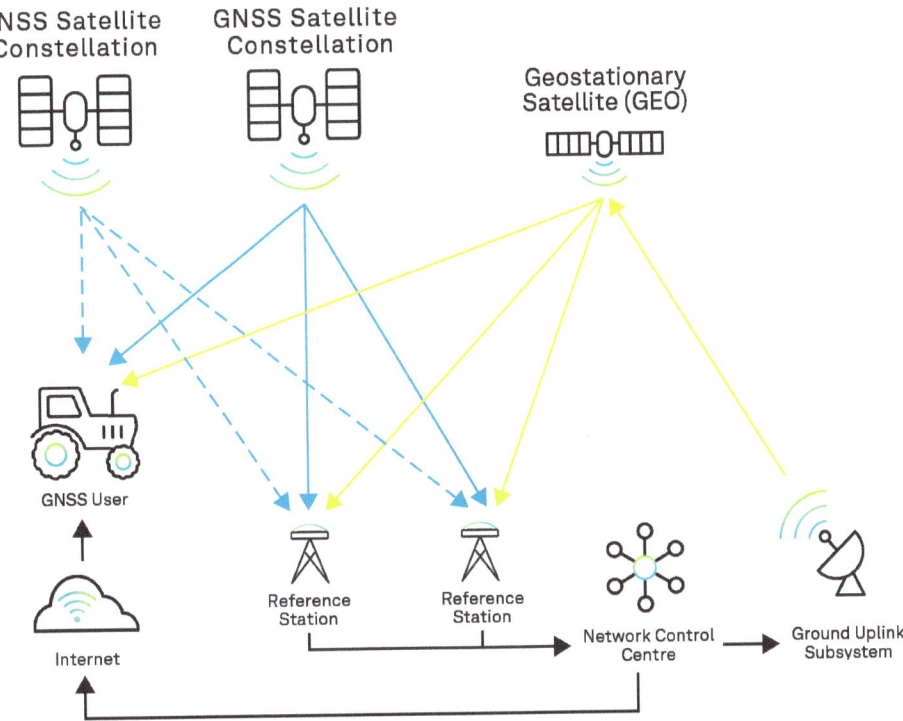

Figure 44 Precise Point Positioning (PPP) system overview

A typical PPP solution requires a period of time to converge to full accuracy in order to resolve any local errors such as atmospheric errors, multipath and receiver biases. The actual accuracy achieved and the convergence time required is dependent on the quality of the corrections and how they are applied in the receiver. Up to 2.5 cm (0.98 inch) accuracy in one minute is possible with advanced corrections and receiver technology.

Similar in structure to an SBAS, a PPP system provides corrections to a receiver to increase position accuracy. However, PPP systems typically provide a greater level of accuracy and charge a fee to access the corrections. PPP systems also allow a single correction stream to be used worldwide, while SBASs are regional. A typical PPP system is shown in **Figure 44**.

The error sources impacting PPP are mitigated by modelling, estimating and applying external corrections.

MODELLING:

Some phenomena are predictable and can be modelled by well-established scientific models. The phenomena that can be well-modelled include the solid Earth tides, antenna phase wind-up and the dry part of the tropospheric delay. The wet part of the tropospheric delay — for example, local humidity levels — can also be modelled, but since it is highly variable, the modelling error must also be estimated.

RESOLVING ERRORS

ESTIMATING:
Unknown components of the system are estimate using conventional estimation strategies. In addition to estimating the receiver's position, many parameters must also be estimated. These include the ionosphere delay, the residual troposphere delay, receiver biases and the carrier phase ambiguities.

APPLYING EXTERNAL CORRECTIONS:
PPP correction providers deliver corrections to account for satellite clock, orbit and biases. In some regions, atmospheric corrections can also be provided in the correction stream. The corrections are typically delivered by satellite or via an Internet connection.

PPP service providers operate a network of ground reference stations to collect correction data for the different signals broadcast by each satellite. The corrections calculated from this data are broadcast from geostationary satellites to the receivers of subscribed users.

GNSS data post-processing

For many applications, such as airborne survey, corrected GNSS positions are not required in real-time. For these applications, raw GNSS satellite measurements are collected and stored for processing post-mission.

During post-processing, base station data can be used from one or more GNSS receivers. Multi-base processing helps preserve high accuracy over large project areas, which is a common occurrence for aerial applications. Depending on the project's proximity to a permanently operating GNSS network, base station data can often be freely downloaded, eliminating the need for establishing your own base station(s). Moreover, it is possible to process without any base station data through PPP, which utilises downloaded precise clock and ephemeris data.

Post-processing applications offer a great deal of flexibility. Applications can involve stationary or moving base stations, and some support integration with customer or third-party software modules. Post-processing applications may be designed to run on personal computers that are accessible through simple-to-use graphical user interfaces.

In the example shown in **Figure 45**, the route taken by the vehicle is shown on the left side of the screen and measurements recorded during the mission, such as velocity, are resolved into horizontal and vertical components in the right side.

Post-processing generally results in a more accurate, comprehensive solution than is possible in real-time.

Which correction method?

As discussed at the start of this chapter, there is no best GNSS correction method, only a method that best suits the intended application. **Figure 46** compares the accuracy and practical range of use for each of the methods discussed in this chapter.

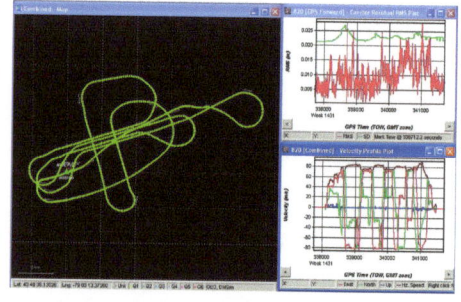

Figure 45 Post-processing of GNSS data

RESOLVING ERRORS

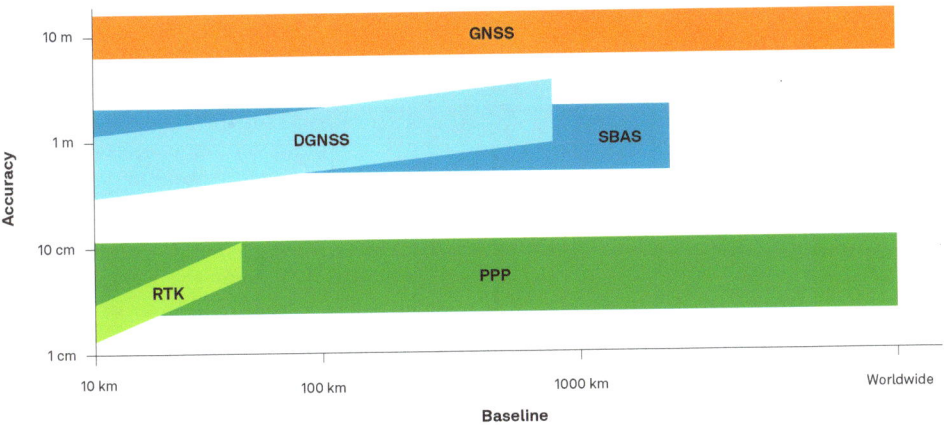

Figure 46 Comparison of GNSS correction method accuracy

The following sections provide comparisons between the correction methods.

DGNSS vs RTK

The configuration of DGNSS and RTK systems are similar in that both methods require a base station receiver set up at a known location, a rover receiver that gets corrections from the base station and a communication link between the two receivers. The difference is that RTK (a carrier phase method) is significantly more accurate than DGNSS (a code-based method).

The advantage of DGNSS is that it is useful over a longer baseline (distance between a base station and rover receivers), and a DGNSS is less expensive. The technology required to achieve the higher accuracy of RTK performance makes the cost of an RTK-capable receiver higher than one that is DGNSS-capable only.

SBAS vs PPP

An SBAS and a PPP system are similar in that both receive corrections from satellites. However, a PPP system is significantly more accurate than an SBAS. Part of the accuracy advantage is the positioning method. PPP uses the carrier phase method, and SBAS uses the code method. The other part of the accuracy advantage is that the private correction services typically used by PPP systems provide higher quality corrections.

The advantage of SBAS is that the correction services are free for everyone to use. While the private correction services provide higher quality corrections and are available worldwide, a paid subscription is required to access the signals.

DGNSS vs SBAS

While the accuracy of DGNSS and SBAS are similar, the equipment required for the system differs.

An SBAS solution only requires an SBAS capable receiver and a GNSS antenna. A DGNSS requires a base station receiver and antenna, a rover receiver and antenna and a communication link between the base station and rover. As well, the DGNSS requires additional system setup

RESOLVING ERRORS

as the base station must be in a known location.

RTK vs PPP

Like DGNSS and SBAS, RTK and PPP offer similar accuracies, but the equipment and setup required is different.

An RTK system typically offers higher accuracy and quick initialisation, but is more complex to setup and more expensive. The RTK system requires at least two RTK capable receivers (one base station and one or more rovers), a GNSS antenna for each receiver and a communication link between the receivers. Also, to achieve the high level of accuracy, the base station must be very precisely set up at a known location.

A PPP solution has a simpler configuration: a single PPP compatible receiver, an antenna capable of receiving GNSS and L-Band frequencies and/ or Internet and a subscription to a corrections service provider. However, PPP has a somewhat lower accuracy and longer initial convergence time, though some advanced PPP systems have reduced these limitations significantly.

Another differentiator is the baseline length. The distance between base station and rover (baseline length) on an RTK system directly impacts system accuracy. At short baseline lengths, a few kilometres or miles, RTK is very accurate. However, as the baseline length increases, the accuracy and availability of a solution decreases. At long baseline lengths, RTK can no longer be used. Because PPP does not use a base station, it is not affected by baseline length and can provide full accuracy anywhere in the world.

Closing remarks

This chapter has described, at a high level, some very complex GNSS concepts around resolving positioning errors. If you want to learn more about these, we have provided a list of references at the end of the book.

58 An Introduction to GNSS, Third Edition

Sensor fusion

In Chapter 5, we described the techniques used to improve GNSS accuracy by reducing the impact of GNSS error sources. In this chapter, we introduce systems in which GNSS receivers work with other sensors to provide positioning and navigation when GNSS conditions are difficult.

A popular term in this field is "sensor fusion." Increasingly it is not just GNSS, or even GNSS + Inertial Navigation System (INS), it is an amalgamation of any and all available information to create the most robust and accurate solution available in all conditions. All the input technologies, GNSS, INS, cameras, odometers, digital elevation models, range sensors, range sensors (LiDAR and Radar), etc. are taken into account.

GNSS+INS systems

As discussed, GNSS uses signals from orbiting satellites to compute position, time and velocity. GNSS navigation has excellent accuracy, provided the antenna has line of sight to at least four satellites. When the line of sight to satellites is blocked by obstructions such as trees or buildings, navigation becomes unreliable or impossible.

An INS uses angular rate and acceleration information from an Inertial Measurement Unit (IMU) to compute a relative position over time. An IMU is made up of six complementary sensors arrayed on three orthogonal axes. On each of the three axes, an accelerometer and a gyroscope are coupled. The accelerometers measure

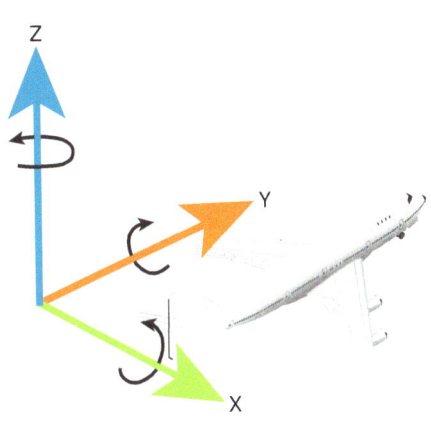

Figure 47 Example of IMU axes

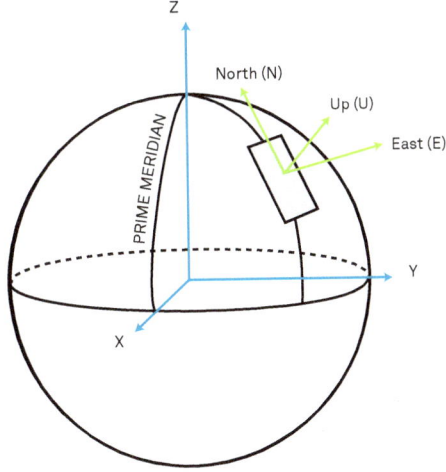

Figure 48 Axes relative to Earth

An Introduction to GNSS, Third Edition

SENSOR FUSION

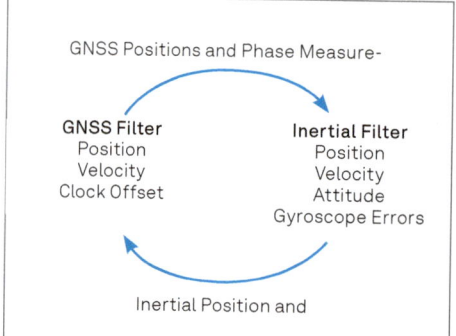

Figure 49 Tightly-coupled GNSS+INS

linear acceleration, and the gyroscopes measure rotational rates. With these sensors, an IMU can measure its precise relative movement in 3D space. The INS uses these measurements to calculate position and velocity. An additional advantage of the IMU measurements is they provide an angular solution about the three axes. The INS translates this angular solution into a local attitude (roll, pitch and azimuth) solution, which it can provide in addition to the position and velocity.

The ability of the INS to provide attitude determination is an important addition for several applications, such as aerial survey, hydrography and autonomous vehicles. For example, in an autonomous vehicle, it is not only important to know where the vehicle is and how fast it is travelling, but also the direction in which the vehicle is headed.

An IMU provides the accelerations and rotations to the INS system as discrete measurements at a specific frequency. Typically, INS systems run at rates between 50 and 1,000 Hz, although most IMUs are capable of sampling their data at much faster rates.

Of course, all systems, including IMUs and therefore INS, have their own drawbacks. First, an INS provides only a relative solution from an initial starting point. This initial start point must be provided to the INS. Second, and more critically, the high-frequency measurements provided by the IMU include several error sources. Depending on the quality (i.e., cost, size) of the IMU, these errors can be fairly large relative to the actual measurements being recorded. Navigating in 3D space with an IMU is effectively a summation (or integration) of hundreds to thousands of samples per second, during which time the errors are also being accumulated. This means that an uncorrected INS will drift from the true position quickly without a way to constrain the error growth. Providing an external reference to the INS, such as a GNSS-derived position, allows it to estimate the errors in the IMU measurements using a mathematical filter and mitigate their effect.

That external reference can quite effectively be provided by GNSS. GNSS provides an absolute set of coordinates that can be used as the initial start point. GNSS provides continuous positions and velocities, which are used to update the INS filter estimates. When GNSS is compromised due to signal obstructions, the INS system can continue to provide effective navigation for longer periods of time.

Using GNSS positions and velocities to estimate INS errors is called a "loosely coupled" system. However, GNSS+INS combined systems can get much more elaborate than that. A variety of terms such as "tightly coupled" or "deeply coupled" clearly indicate a much more symbiotic relationship between the two. In these systems, raw GNSS measurements are used directly to aid

SENSOR FUSION

the INS, and the INS can even be used as a constraint to help GNSS reacquire lost signals more quickly or reject bad signals. **Figure 49** shows a simplified diagram of a tightly coupled system.

Thus, when GNSS and INS are combined, the two techniques enhance each other to provide a powerful navigation solution, as illustrated in **Figure 50**. When the GNSS conditions are good (line of sight to several satellites), the GNSS receiver provides accurate position and time to the navigation system. When the GNSS conditions become poor, the INS provides the position and navigation until the GNSS conditions improve.

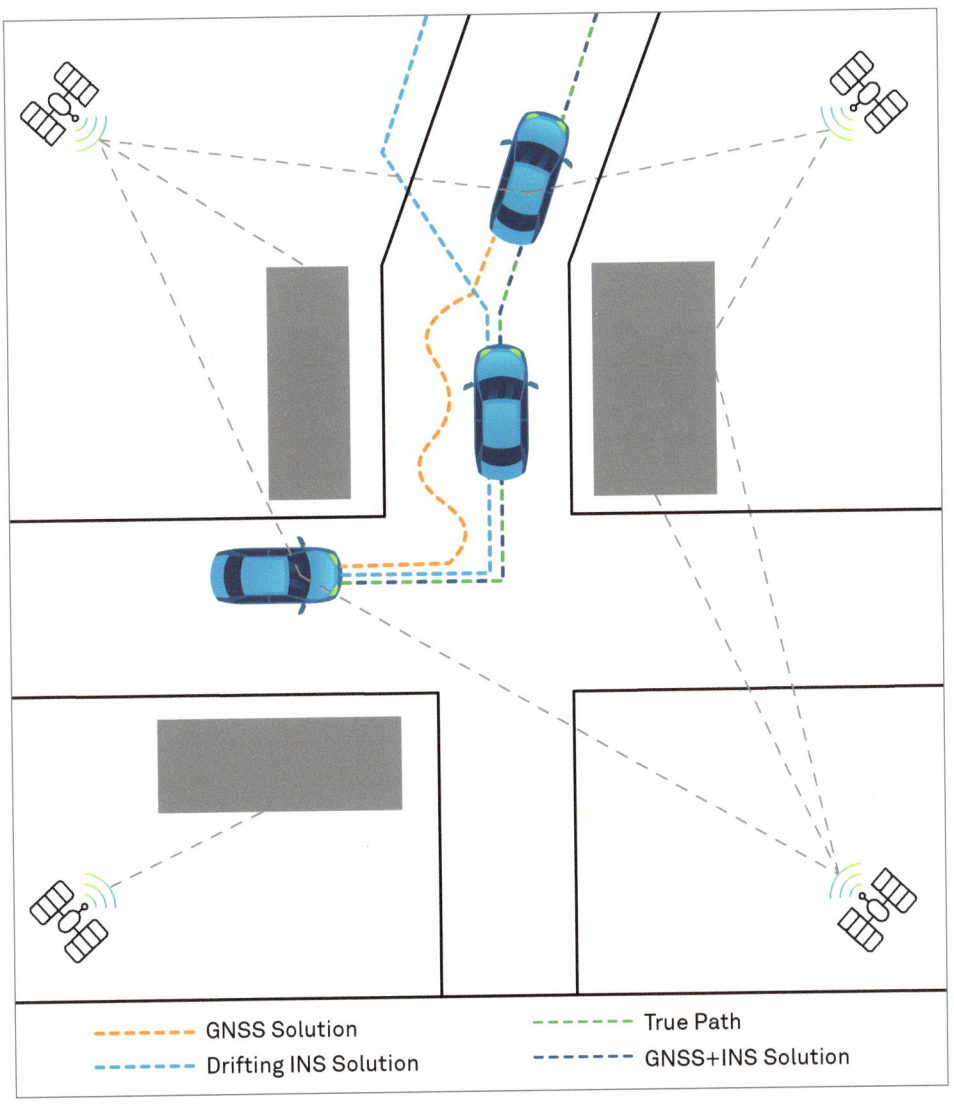

Figure 50 Combined GNSS+INS solu-

SENSOR FUSION

Odometers

GNSS is not the only useful input to aid inertial navigation. For different environments, different sensors can also be added to aid the solution. A common external sensor for ground vehicles is the addition of a Distance Measurement Instrument (DMI) such as an odometer. This provides another independent measurement of displacement and velocity that can aid the GNSS+INS navigation solution. This is mainly of use when the GNSS signal is denied, for example, when travelling through a tunnel.

Vision-aided navigation

Another potential aiding source is the use of photogrammetry or using vision-aided navigation. In a vision-aided navigation system, imagery is used to provide position information to a navigation system. Images from a camera are processed by the navigation system to recognise and track objects in the environment.

There are two ways this can be used. Known surveyed camera targets can be used to generate an absolute position in a certain environment or everyday objects can be used as control point. When an object is recognised by the system, the relative change in successive images can be used to generate a relative position change of the camera in 3D space.

This means that a vision-aided system can be combined with a GNSS+INS system to provide position and attitude updates to the INS when GNSS updates are not available. An example application for a vision-aided navigation system is an autonomous vehicle used to carry a load from a yard into a warehouse. When the autonomous vehicle is outside, the GNSS+INS provides the navigation for the vehicle. When inside the warehouse,

64   An Introduction to GNSS, Third Edition

the vision-aided system uses known features/targets within the building to provide position updates to the INS.

Closing remarks

If you want to learn more about the topic in this chapter, we have provided a list of references at the end of the book.

GNSS threats

When GNSS signals are jammed (even unintentionally), spoofed or blocked by obstacles (such as urban canyons or dense foliage), a GNSS receiver may not be able to provide positioning, navigation and timing (PNT) data. The following sections describe the different ways by which GNSS signals can be threatened, and various methods to mitigate such threats.

Interference

By the time GNSS signals arrive at the antennas of a GNSS positioning system, the power level of these signals is very low (nearly 1 quintillion times weaker than a typical light bulb). This low power level makes the signals susceptible to interference from other signals transmitted in the GNSS frequency range. If the interfering signal is sufficiently

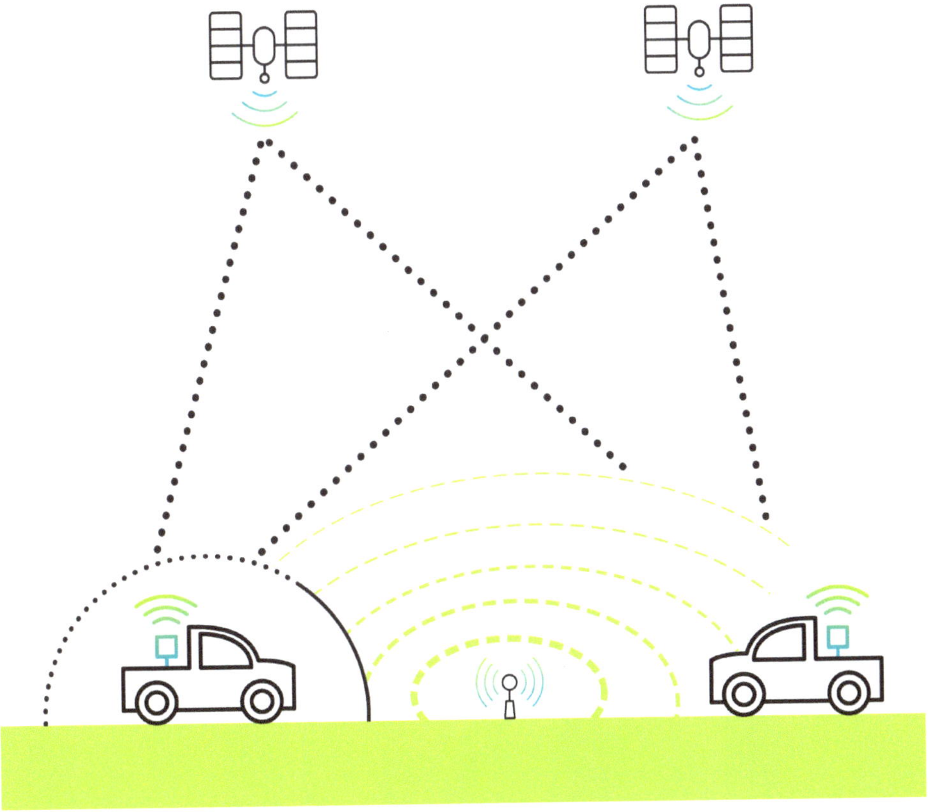

Figure 51 Anti-jam antenna protecting against a jammer

GNSS THREATS

powerful, it becomes impossible for the receiver to detect the low-power GNSS signal.

If the signal is from an unintentional source, it is generally called interference. There have been many documented cases of unintentional interference over the years, ranging from faulty TV receivers to other non-GNSS transmitting sources leaking into the GNSS frequency bands.

Jamming occurs when a signal is broadcast within the GNSS band with the intention to overwhelm GNSS signals and prevent receivers in the area from providing PNT data. Hostile entities may try to jam GNSS signals in a local area or region for a number of reasons, with the ultimate intent to make GNSS signals unusable. Although illegal in most jurisdictions, very low-power jammers known as "personal privacy devices" can be bought on the Internet and used for this purpose. One simple, low-power jammer can overpower GNSS signals within a large geographic area, denying a position solution and timing.

Another version of interference (the above is in-band interference) is band-adjacent. Some jammers use high-powered transmitters to overwhelm the bands adjacent to GNSS bands and compromise a receiver's ability to receive GNSS signals.

GNSS receivers can use several methods to protect against interference and jamming.

Signal filtering

The first line of defence for interference in any radio frequency (RF) system is to filter out as much of the interference as possible. High-quality GNSS antennas provide the first layer of filtering as they are designed to increase the signal gain in the GNSS band and attenuate signals that are out-of-band. GNSS receivers also have filters that reduce the signal power of out-of-band signals which helps against band-adjecent jamming. Some advanced GNSS receivers include tools that detect interfering signals and create filters that can reduce both in-band and out-of-band interference.

Multiple navigation sensors

For short-term interference, additional navigation sensors, such as Inertial Measurement Units (IMUs), odometers or altimeters, can help the receiver bridge brief periods of GNSS outage. A discussion of systems that use GNSS receivers and IMUs, called GNSS+Inertial Navigation Systems (INS), is presented in Chapter 6.

Multi-frequency/multi-constellation GNSS

Multi-frequency receivers also provide a means of protecting against interference. For example, if interference in the L1 frequency band around 1575 MHz is completely jamming GPS L1, GLONASS L1, Galileo E1 and BeiDou B1, it is usually possible to continue calculating a positioning, navigation and timing (PNT) solution using GPS L2 and GLONASS L2 around 1227 MHz or GPS L5, Galileo E5a and BeiDou B2a around 1176 MHz.

Additionally, some of the newer and faster civilian signals such as GPS L5, Galileo E5a and Galileo E5b provide some improvement in interference

GNSS THREATS

performance.

Encrypted signals (such as the Y-code and M-code signals on GPS L1 and L2) broadcast by several of the GNSS constellations provide significant additional resistance to interference and jamming.

A description of GNSS signals is given in Chapter 3.

Anti-jam antennas

Anti-jam antenna systems, comprising Controlled Reception Pattern Antennas (CRPAs) and sophisticated electronics, use multiple antenna elements to control the amount of signal received from a particular direction. When an anti-jam system senses interference from one direction, it turns down the antenna gain, thereby reducing the received signal power for that direction. This process reduces the amount of interference received so that legitimate GNSS signals can be received from other directions.

Figure 51 shows two vehicles in range of a GNSS jammer. The vehicle on the right has a standard antenna, and the GNSS signals are overpowered by the jammer. The vehicle on the left has an anti-jam antenna that blocks the jamming signal so GNSS signals can be received.

Spoofing

Unlike interference where GNSS is denied by overpowering the satellite signal, spoofing tricks the receiver into reporting an incorrect position. Spoofing is done by first jamming the GNSS receiver then providing a false satellite signal that is either created by a signal generator or is a rebroadcast of a real recorded GNSS signal. Unlike interference, spoofing is always an intentional attack.

To deny GNSS by spoofing, the attacker broadcasts a signal with the same structure and frequency as the GNSS signal. The spoofing signal controls its transmitted power level, so the receiver will lock onto the spoofed signal rather than the real GNSS signal. In the spoofed signal, the message is changed so that the receiver will calculate an incorrect position or time.

The most effective way to protect against spoofing is to track encrypted signals (such as the Y-code and M-code signals on GPS L1 and L2) that are broadcast by several of the GNSS constellations. Access to the encrypted signals is restricted and not available to all users; however, there are mitigation methods that can be used with open signal receivers.

The complexity of spoofing increases greatly if the attacker attempts to simultaneously spoof more than one GNSS frequency or constellation. So, a receiver that can track multiple frequencies and/or multiple constellations can be used to detect and overcome a possible spoofing attempt.

Other navigation sensors, such as GNSS+INS, can be used to detect and overcome a spoofing attempt as the measurements from the IMU cannot be spoofed.

Signal blockage

A GNSS receiver needs a clear line of sight to the satellites it is tracking. If the line of sight to a satellite is blocked by objects such as buildings, trees, bridges,

GNSS THREATS

etc., the receiver cannot receive signals from that satellite. In locations that have a lot of obstructions, such as the urban canyons found in downtown areas of a large city, the obstructions can block so many satellites that the receiver cannot calculate its position or time.

One solution to signal blockage is for the receiver to track more than one constellation. By tracking more than one constellation, there will be more satellites available and a better chance of finding enough satellites to determine a position and time. The use of multiple navigation sensors, such as IMUs, helps not only in bridging outages such as those due to signal blockage, but also in the reacquisition of the GNSS signals after the outage.

Constellation failure

Although it is extremely unlikely that an entire constellation will fail, receivers that can track more than one constellation provide protection from this unlikely scenario.

Closing remarks

With the many technologies and applications depending on GNSS to provide PNT, the topic of protecting GNSS is becoming increasingly important. If you want to learn more about this subject, we have provided a list of references at the end of the book.

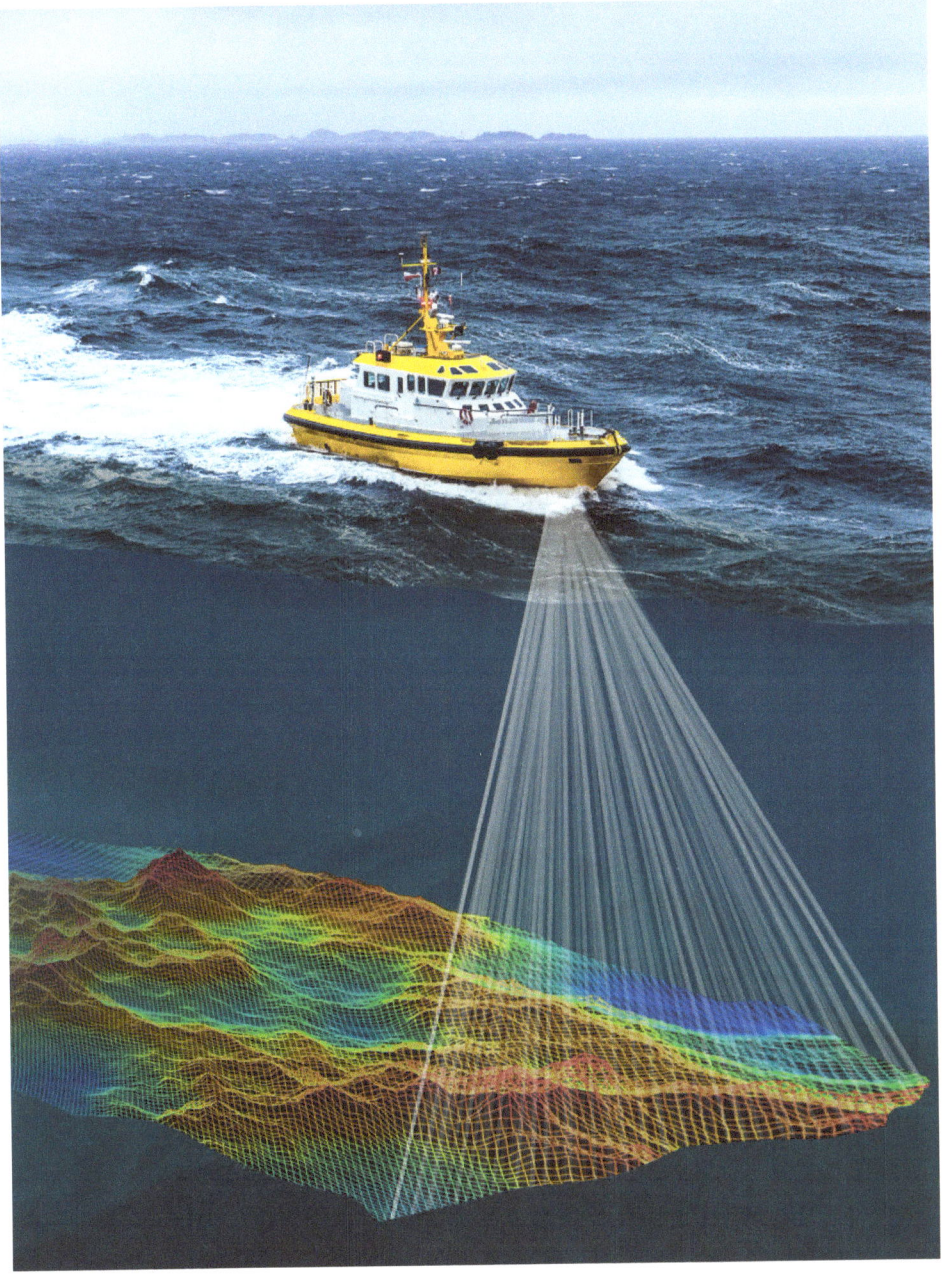

"I think the age of exploration is just beginning, not ending, on our planet."
–Robert Ballard

AUTONOMY

Autonomy

We have built an understanding of the concepts behind positioning. This chapter will explore some of the technologies behind Autonomous Vehicles (AVs) and how positioning can help replace human error with the ability to navigate safely at all times.

Autonomy levels

Autonomous features are increasingly being added to production vehicles, taking us closer to the emergence of fully autonomous road, sky and sea navigation. By understanding what capabilities an entirely self-driving vehicle would have compared to a vehicle with no AV characteristics, we can better understand the direction of AV technology.

An industry standard developed by the Society of Automotive Engineers (SAE) provides guidance on differing autonomy levels. This standard, defining six stages of vehicle automation, from levels zero to five, offers a helpful baseline for referencing what stage of autonomy a particular vehicle is (**Figure 52**). Level 0 vehicles are entirely human-dependent, with the driver performing all tasks related to positioning and safety. Vehicles with Advanced Driver Assistance System (ADAS) technologies such as Adaptive Cruise Control and Parking Sensors can provide some Level 1 or 2 assistance. Increasingly advanced, efficiently integrated technologies result in fully automated vehicles at Level 5, with no requirement for ongoing human interaction.

Sensors

AVs contain numerous dedicated devices known as sensors to assist with self-driving navigation. A sensor monitors the physical environment, responding to different stimuli by producing different electrical signals.

AVs cannot rely on one individual sensor to visualise the surrounding environment. Sensors operate and deliver data in different ways, and each sensor has its advantages and disadvantages. However,

LEVEL 0

No Automation

Driver performs all tasks related to driving.

LEVEL 1

Driver Assistance

Driver performs driving tasks with assistance from built-in safety features. The driver maintains control of the vehicle.

LEVEL 2

Partial Automation

Driver must remain alert, engaged in the environment and in control. Vehicle can perform one or more tasks simultaneously.

LEVEL 3

Conditional Automation

The vehicle monitors the environment and can perform driving tasks. Driver is not required to remain alert but must be ready to take control.

LEVEL 4

High Automation

Vehicle performs all driving tasks and monitors environment under limited conditions. Driver may still be required to intervene.

LEVEL 5

Full Automation

Vehicle performs all driving functions under all conditions. Driver is not required.

Figure 52 Autonomy levels

LiDAR

Inside a LiDAR (light detection and ranging) device are multiple stacked semiconductors that fire lasers (tightly focused beams of photons). These semiconductors can be rotor mounted to provide a 360° field of view or, more commonly, solid-state and mounted to fire arrays over a specific field of view.

Once a laser hits an object, it returns to be picked up by an onboard infrared detector. The duration from the emission of a laser to its return (time-of-flight) determines how far away an object is. Poor weather can impede the capability of LiDAR, with fog or rain reducing the integrity of the data LiDAR can produce.

Radar

Also operating on the time-of-flight principle but less weather-dependent is radar (radio detection and ranging), which emits radio waves rather than lasers. Radar sensors are typically fixed, focusing radio waves in one direction. Although radar usually results in a lower resolution picture than LiDAR, radio waves have a longer wavelength than light, meaning they can travel further and thus detect objects at a greater distance.

Noticing shifts in the radio wave frequency of a detected object can also help determine if it is moving and its relative speed. As an object comes closer to the radar, the reflected frequency will increase as the radio waves become more compressed; as an object moves away, the frequency will decrease as the radio waves flatten out.

While often mounted onto the front, sides and rear of a vehicle, the ability to detect objects at a distance and determine relative movement makes front-facing radar a particularly compelling choice for forward-collision warning, avoidance and adaptive cruise control.

Cameras

We have explored how an AV can map surroundings with LiDAR and radar alone to determine objects' location and relative movement. However, identifying the surrounding objects and determining where the AV is remains restricted without additional input. Cameras can provide an AV with the means to recognise and track objects in the environment and basic positioning.

When objects are identified and used as control points, the change in successive images captured can provide relative positioning as the AV travels through 3D space. By helping identify known surveyed targets such as landmarks, cameras can also generate an absolute position in an environment. The relative usefulness of a camera is dependent on the availability of visible objects. Featureless surroundings or environments severely impeded by poor lighting or local weather can hamper camera effectiveness.

GNSS and INS integration

We can expect unaided GNSS accuracy to be about 2 to 5 metres (2.2 to 5.5 yards) 95% of the time. The demands for self-driving vehicles are much more stringent than this. AVs are typically looking for accuracy in the order of 2 decimetres (7.8 inches) and need guarantees that the error is smaller than the width of a standard lane.

Achieving this increased level of accuracy requires a source of external corrections to allow carrier phase positioning. The industry has more or less settled on PPP due to its advantages over RTK in large-scale deployments. GNSS complements vision and localisation

AUTONOMY

Figure 53 GNSS capability in different environments

sensors by providing a source of absolute positioning, independent from relative positioning sensors. GNSS is useful in open-sky environments (**Figure 53**, top panel) where there are few landmarks or objects for sensors to reference for relative positioning, and during adverse weather conditions (**Figure 53**, middle panel), which don't pose a problem for GNSS, as it is almost entirely immune to weather effects.

PPP limitations become apparent when driving under an overpass, for example (**Figure 53**, lower panel). PPP requires a continuous view of every satellite in use, and as soon as that's broken, PPP is no longer available and will need time to reconverge. Such environments require additional sensor input to aid the GNSS, with IMUs being the most obvious addition. An IMU calculating relative motion by measuring 3D acceleration and rotation can generate data to help provide positioning, velocity and attitude throughout a PPP outage and reconvergence period.

Sensor fusion

Each sensor has its purposes, strengths and weaknesses, and no one sensor can be solely relied upon to create a safe, fully autonomous vehicle. The use of software algorithms to verify the output integrity of multiple sensors and tie valid outputs together with GNSS and INS to create a high-definition environmental map is known as sensor fusion. This magic mix combines the sum of all parts to allow for the exponential increase in reliability that occurs when amalgamating multiple data outputs.

It's not only sensors dedicated to

AUTONOMY

autonomy such as LiDAR, radar and cameras that are useful in sensor fusion. Most modern commercial and industrial road-based vehicles contain many other useful sensors, most coming from dynamic stability control systems or electronic stability control. Such sensors have existed for a long time but became mandatory in the U.S. in 2012 and the EU in 2014.

In a modern vehicle, you can expect to find speed sensors on every wheel, yaw rate sensors, steering angle sensors, transmission settings, throttle and brake sensors, all transmitting information to a highspeed data bus (**Figure 54**). Combining these sensor outputs alone with an onboard IMU can provide a positioning solution with redundant measurements for velocity, turn rate and vehicle direction.

While GNSS provides the only real-time absolute positioning solution, it is prone to loss. Unlike GNSS, IMU data is always available and can help provide vital relative positioning during a GNSS loss. The problem with INS is that it measures acceleration and rotation, not position, meaning the integration of acceleration and rotation must occur once to get velocity and again to get a position. Any acceleration or rotation rate measurement error will lead to an exponential growth in position errors. Additionally, like any micro-electromechanical systems (MEMS) sensor, an IMU is prone to white noise and drift, introducing additional errors. Thankfully, sensor fusion can use the output from other sensors to constrain error growth and make a substantial difference in GNSS outage bridging.

Functional safety

Autonomous applications, by definition, are designed to supersede human oversight. With no human to help determine if there is an error, it is vital sensor outputs used by sensor fusion in autonomous applications are reliable. Using an erroneous position solution, an AV may make decisions based on incorrect data, leading to potentially hazardous situations. Before sensor fusion can combine an output source with other data for positioning, it requires a guarantee of integrity. AVs can only be expected to be widely adopted when the underpinning technology can be proven to be consistently reliable.

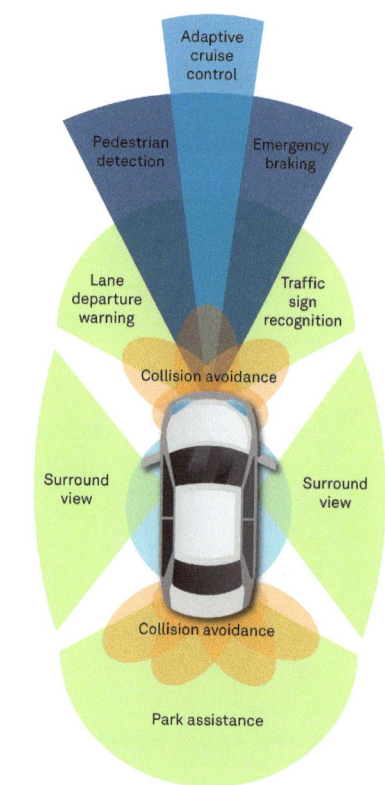

Figure 54 Sensor applications

Traditional positioning safety

How positioning data is validated is crucial to ensuring integrity. Standard deviation is the accuracy metric normally used for positioning, which measures how far each output value is from the mean and specifies the expected error level in the output value.

We can typically be more confident in positioning output datasets with a small standard deviation, where the positioning values are tightly clustered together and with relatively few outliers (**Figure 55**). When using a GNSS receiver to determine positioning, it is ideal to have many well-spread satellites available, as this helps provide greater accuracy. Latitude and longitude values for a fixed location, derived from healthy satellite geometry, will usually be centred when represented on a scatter plot and result in a small standard deviation.

Positioning is simply an estimation problem, and all measurements have errors, typically distributed in a centred or "Gaussian" (bell curve) manner. Standard deviation helps us quantify accuracy expectations from our positioning errors and determine when normal, or Gaussian, data value distributions are present. The use of standard deviation to assign a confidence ellipse helps define the likelihood of an actual position falling within a designated region a certain percentage of the time. For example, as portrayed in green in **Figure 55**, a 3-sigma confidence ellipse (3 standard deviations) means that we can expect the actual position to fall within this defined area 99.7% of the time.

As standard deviation measures how far each output value is from the mean, you might think that simply widening the error ellipse would help to ensure integrity. Unfortunately, this breaks down when measurements are no longer Gaussian or have a systematic bias due to measurement faults. The standard deviation approach does not mitigate against positioning errors such as multipath (as discussed in Chapter 4), represented in **Figure 56** by dark blue dots.

Beyond multipath, the GNSS receiver alone may encounter many other issues such as satellite orbit calculation error or ionospheric interference. However, in designing an AV positioning system, every potential failure point for every sensor needs to be considered for all possible operating conditions, which equates to billions of possible scenarios where a

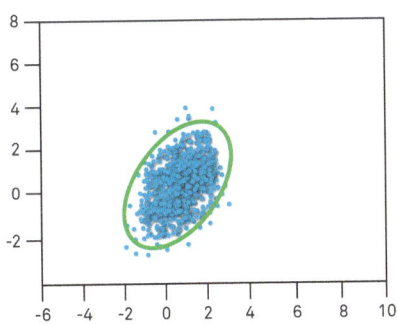

Figure 55 Output data (blue) and 3-sigma confidence ellipse (green)

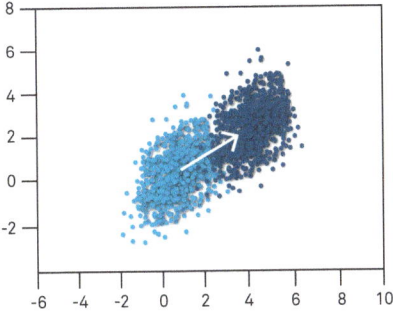

Figure 56 Multipath error (dark blue)

AUTONOMY

failure could occur. The use of a protection level supersedes the confidence ellipse for autonomous applications that require high integrity inputs.

Autonomous positioning safety

Protection levels provide a means to quantify and guarantee the maximum degree of error expected to exist by calculating a robust maximum error bound against erroneous input data. Looking at an example, suppose that our GNSS receiver has allowed us to determine a 3-sigma position error accurate to within 2 metres, corresponding to the blue sphere in **Figure 57**. The GNSS receiver may also separately provide us with a 99.9999% protection level of 8 metres (8.8 yards). The GNSS receiver has effectively calculated a 0.0001% chance that our actual position falls outside the 8 metres (8.8 yards) green zone of **Figure 57**.

Protection levels will grow and shrink dynamically to maintain the selected confidence level in response to different environments. A drop in satellite visibility introduced to our data in **Figure 57** may result in the protection level increasing from 8 metres (8.8 yards) to 10 metres (10.9 yards) to maintain the 99.9999% confidence guarantee, as there is now a higher chance that the error could be greater than 8 metres (8.8 yards) but less than 10 metres (10.9 yards). The protection level may grow to a point where it exceeds a static reliability threshold known as an alert limit and is no longer considered safe.

Figure 58 shows a Stanford diagram, which provides a more precise means to visualise the relationship between positioning errors, protection levels and alert limits. **Figure 57** provided an example of a protection level that grew in response to errors and remained greater than the positioning error. As the protection level tries to account for the worst possible undetected faults, protection levels will normally remain greater than positioning errors. Consequently, as a Stanford diagram plots position error against protection level, every position point in the diagram should fall within the normal operation zone, as shown in **Figure 58**.

An alert limit of 3 metres (3.3 yards) is in place to ensure the use of the position only if the error is guaranteed to be less than 3 metres (3.3 yards). The alert limit would trigger in response to a protection level greater than 3 metres (3.3 yards), rejecting the position from use as it would no longer be considered safe. Misleading

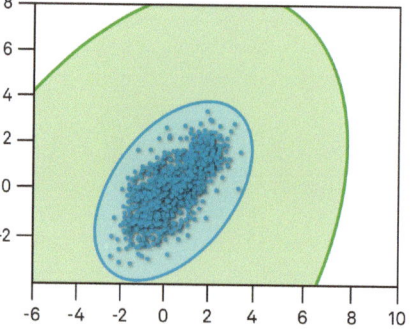

Figure 57 Protection level area shown in green

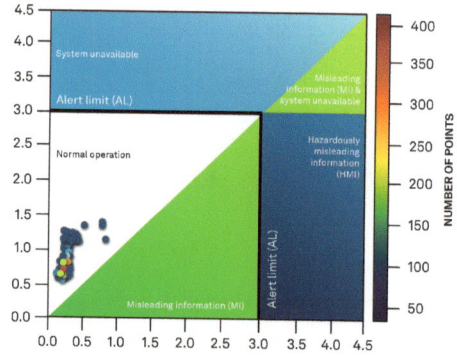

Figure 58 Stanford diagram

AUTONOMY

information occurs when position errors are greater than the protection level but below the alert limit. Such errors are not flagged as hazardous since the protection level remains less than the alert limit, but they are misleading since the error exceeds the protection level. It is crucial in safety- critical applications to avoid the worst-case scenario, where the position error exceeds the alert limit, but the protection level remains below the alert limit. This type of error would be hazardously misleading information since the protection level tells the user that the position is safe to use when the error exceeds the alert limit.

We have used the idea of the GNSS receiver calculating protection levels. In reality, a separate system may handle calculating protection levels for the GNSS receiver and other sensors in a process known as centralised or parallel processing. While perfect for helping visualise the surrounding environment, most AV cameras, for instance, may have insufficient processing power to determine protection levels. The use of dedicated equipment allows a place to run the advanced algorithms necessary to establish protection levels and perform the broader task of sensor fusion, tying together the various sensor inputs. Centralised processing also helps keep the design of an AV cost-effective and scalable in production while guaranteeing the protection levels necessary to aid reliable autonomous navigation.

Safety-critical design

Functional safety is the fine art of quantifying the extent of trust held in a system to perform safely in hazardous conditions and ensuring that mitigating actions are in place should an error occur. AV development must consider all possible errors and meet required automotive safety standards, such as ISO 26262 for on-road applications, from initial concept to final validation and testing. Addressing thousands of potential failure points, it may be no surprise that determining integrity throughout the entire development cycle can easily become a massive undertaking.

We have focused on how we can establish the absolute integrity of an AV without consideration for how the various underlying AV technologies communicate. Vehicles such as cars built with AV technology allow sensors to communicate via the internal data buses found on most modern automobiles. Acting like the nervous system of a vehicle, the communication bus provides a robust and reliable primary means for onboard sensors to communicate and aid navigation. However, AVs also incorporate increasingly externally connected technology, which leads to more significant potential for accidental user misuse of the system or an intentional malicious interception of a vehicle remotely. Vehicle safety may become compromised by a user interacting with the vehicle in an unanticipated manner or an attacker exploiting any number of available attack vectors. The ISO 21434 Road vehicles — Cybersecurity engineering standard provides a guideline to manage potential cybersecurity threats during the development process.

When it comes to building a safe and more secure AV system, from initial concept to product delivery, well-defined processes must be in place to meet stringent safety and security standards. The most effective approaches consider both functional safety and cybersecurity considerations efficiently and holistically.

AUTONOMY

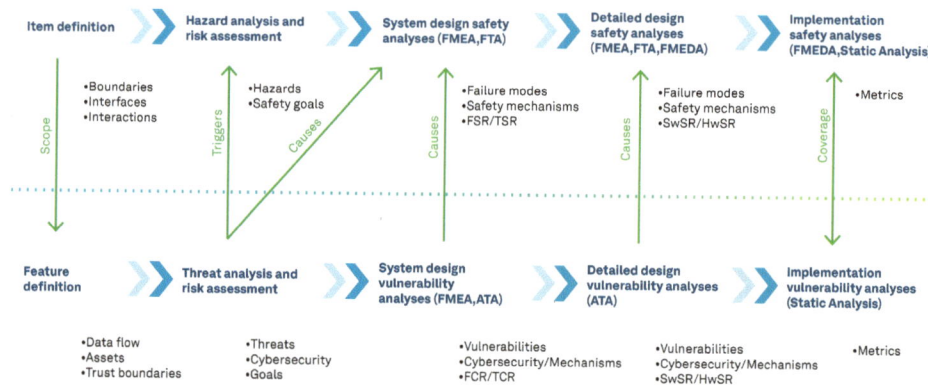

Figure 59 Functional safety and cybersecurity analysis streams

One such approach during the development lifecycle is to merge both the functional safety and cybersecurity analysis streams. **Figure 59** shows an overview of both functional safety and cybersecurity analysis streams. Separately they are sufficient to meet the requirements set out in ISO 26262 and ISO 21434; however, a more effective development lifecycle is possible when executed in parallel, with avenues for integration implemented.

Upon closer inspection, we can see that the initial functional safety analysis activity is Item Definition, which details boundaries, interfaces and interactions. Once complete, this activity is available as an input to the initial cybersecurity Feature Definition step, which defines the flow of data, the assets and the trust levels between different elements and informs the cybersecurity Threat Analysis and Risk Assessment activity. This step can provide vital considerations to the functional safety Hazard Analysis and Risk Assessment activity, as any compromised component could potentially cause malfunctioning behaviour in multiple other system components or software. These initial steps provide a means to formulate safety goals that the rest of the development cycle must fulfil to determine failure modes, safety mechanisms and the metrics for testing functional safety and cybersecurity requirements.

Closing remarks

You should now feel more familiar with the solutions incorporated into making autonomous, inherently safe vehicles a reality. If you want to learn more about this topic, we have provided a list of references at the end of the book.

AUTONOMY

GNSS APPLICATIONS AND EQUIPMENT

GNSS applications and equipment

"Every generation needs a new revolution."
—**Thomas Jefferson**, influential Founding Father and third president of the United States.

We don't think it is an overstatement to say that GNSS technologies have revolutionised, and continue to revolutionise, the way businesses and governments operate. This chapter highlights some of the incredible GNSS applications and equipment that are now available.

Applications

In a short book, it is impossible to describe all GNSS applications. We will highlight some of the commercial applications, including:

- Consumer
- Transportation
- Machine control
- Precision agriculture
- Construction
- Surface mining
- Survey
- Aerial photogrammetry
- Ground mapping
- Geospatial information systems (GIS)
- Timing
- Marine
- Unmanned vehicles
- Defence

Consumer

GNSS technology has been adopted by the consumer market, in an ever-increasing range of products.

GNSS receivers are now routinely integrated into smartphones to support applications that display maps showing the location of and the best route to stores and restaurants.

Portable navigation devices give drivers directions on-road or off, as shown in **Figure 60**.

Geocaching is an outdoor activity in which participants use a GNSS receiver to hide and seek containers (called "geocaches" or "caches") around the world.

Transportation

"I knew I was going to take the wrong train, so I left early."
—Yogi Berra, former Major League Baseball player and manager.

In rail transportation, GNSS is used in conjunction with other technologies to track the location of locomotives and rail cars, maintenance vehicles and wayside equipment for display at central monitoring consoles. Knowing the precise location of rail equipment reduces accidents, delays and operating costs, enhancing safety, track capacity and customer service.

In aviation, GNSS is being used for aircraft navigation for departure, en route

Figure 60 Portable

An Introduction to GNSS, Third Edition 83

GNSS APPLICATIONS AND EQUIPMENT

and landing. GNSS facilitates aircraft navigation in remote areas that are not well served by ground-based navigation aids, and it is a significant component of collision-avoidance systems, and systems used to improve approaches to airport runways. Refer to Chapter 5 for information about WAAS, a regionally based system across North America that delivers GPS corrections and a certified level of integrity to the aviation industry, enabling aircraft to conduct precision approaches to airports.

In marine transportation, GNSS is being used to accurately determine the position of ships when they are in the open sea and when they are manoeuvring in congested ports. GNSS is incorporated into underwater surveying, buoy positioning, navigation hazard location, dredging and mapping.

In surface transportation, vehicle location and in-vehicle navigation systems are being used throughout the world. Many vehicles are equipped with navigation displays that superimpose vehicle location and status on maps. GNSS is used in systems that track and forecast the movement of freight and monitor road networks, improving efficiency and enhancing driver safety.

Port automation

Using GNSS, shipping hubs can improve their operating efficiency by tracking the movement and placement of containers throughout their yards.

Gantry cranes are used in ports throughout the world to lift shipping containers, as shown in **Figure 61**. These cranes are large and sometimes difficult to steer accurately in a crowded shipping dock. Many cranes are equipped with GNSS-based steering devices that determine the crane's position and keep it travelling in the desired path, improving accuracy and productivity as well as the safety of operators and workers on the ground. A key benefit is the quick movement of containers about the port, which reduces food spoilage and gets freight delivered on time.

Autonomous driving

One of the most exciting applications of GNSS is enabling autonomous driving — both on-road and off-road. As discussed in Chapter 8, the Society of Automotive Engineers (SAE) has defined levels of vehicle automation from 0 to 5 (**Figure 52**). Above Level 3, the vehicle monitors the environment and performs driving tasks. The positioning accuracy, reliability and integrity needed to enable safety-critical autonomy requires a combination of high-precision GNSS technologies, sensors and corrections. A few of the initial applications of autonomous vehicles

Figure 61 Gantry crane for moving shipping containers

GNSS APPLICATIONS AND EQUIPMENT

include university campus shuttles, last-mile product delivery and taxis. Several universities are engaged in research to develop autonomous vehicle platforms. One example is the University of Iowa's Automated Driving Systems (ADS) for Rural America project, which is now driving a partially automated shuttle bus through rural Iowa to enhance mobility for transportation-challenged populations such as the aging in rural communities.

An article (On the Rural Road to Autonomy) about the ADS for Rural America is in the 2022 Velocity magazine available at: resources.hexagonpositioning.com/rural-road-autonomy.

Parking automation

In the Canadian city of Calgary, paying for on-street parking has become automated. Customers pay for parking at street side terminals or using their smartphone, and monitoring of the parked vehicles is done from a vehicle equipped with cameras and a GNSS receiver.

When a customer pays for parking, they enter the licence plate of their vehicle, a code that identifies the parking area and the amount of parking time required. This information is sent to a database. As the monitoring vehicle drives along the street, the vehicle cameras capture the licence plates of the parked cars. The licence plate number, along with the time and position provided by the GNSS receiver, is compared to the database of paid parking. If a vehicle is not found in the database, the photograph is sent to a Calgary Parking Authority employee so they can determine if there is a just cause (e.g., people are just getting out of the car), there is a mistake in identifying the licence plate (e.g., a D is mistaken for an O) or if it is a parking violation.

Due to the tight urban corridors in downtown Calgary, the location reported by the GNSS-only system on the monitoring vehicles was misplacing 6-7% of the vehicles (approximately 1,400 vehicles) per day and the vehicles could be misplaced by up to 600 metres (657 yards). This misplacement caused many hours of extra work each day for Calgary Parking Authority employees, because they had to manually correct the vehicle's position before they could determine if there was a parking violation.

By switching from a GNSS-only system to a GNSS+INS system, the monitoring vehicle was able to overcome the GNSS challenges in downtown Calgary and provide a much more reliable position. The GNSS+INS system reduced the number of misplaced cars to 1% (less than 300 vehicles), saving the Calgary Parking Authority enough work hours to pay for the GNSS+INS system in all their monitoring vehicles in under two years.

An article (Calgary ParkPlus Program, City-Wide Positional Accuracy) about how GNSS+INS has helped the Calgary Parking Authority is in the 2014 Velocity magazine available at: resources.hexagonpositioning.com/calgary-parkplus-program.

Guiding snowplows in the Andes

The high mountain roads in the Andes mountains of Argentina provide important links for local communities, cross-border traffic and trade. During the winter months, these snowbound roads are difficult to navigate and sometimes dangerous. Even the vehicles and machinery used to keep these routes open need help to navigate when snowdrifts make the roadways difficult to follow.

For the snowplows keeping these roads clear, a GNSS receiver with PPP positioning technology is used to ensure the snowplow stays on the road even when the driver cannot see the road or signs.

An article (MicroElect takes positioning into the Andes Mountains) about how GNSS has helped snowplows in Argentina is in the 2019 Velocity magazine available at: resources.hexagonpositioning.com/microelect-positioning-andes-mountains.

Machine control

GNSS technology is being integrated into equipment such as bulldozers, excavators, graders, pavers and farm machinery to enhance productivity in the real-time operation of this equipment, and to provide situational awareness information to the equipment operator. The adoption of GNSS-based machine control is similar in its impact to the earlier adoption of hydraulics technology in machinery, which has had a profound effect on productivity and reliability.

Some of the benefits of GNSS-based machine control are summarised below:

EFFICIENCY:
By helping the equipment operator get to the desired grade more quickly, GNSS helps speed up the work, reducing capital and operating costs.

ACCURACY:
The precision achievable by GNSS-based solutions minimises the need to stop work while a survey crew measures the grade.

JOB MANAGEMENT:
Managers and contractors have access to accurate information about the job site, and the information can be viewed remotely.

DATA MANAGEMENT:
Users can print out status reports, save important data and transfer files to head office.

THEFT DETECTION:
GNSS allows users to define a "virtual fence" around their equipment and property for the purpose of automatically raising an alarm when equipment is removed, then providing equipment tracking information to the authorities.

Precision agriculture

In precision agriculture, GNSS-based applications are used to support farm planning, field mapping, soil sampling, tractor guidance and crop assessment. More precise application of fertilisers, pesticides and herbicides reduces cost and environmental impact. GNSS applications can automatically guide farm implements along the contours of the Earth in a manner that controls erosion and maximises the effectiveness of irrigation systems. Farm machinery can be operated at higher speeds, day and night, with increased accuracy. This increased accuracy saves time and fuel, and maximises the efficiency of the operation. Operator safety is also increased by greatly reducing fatigue.

Figure 62 Automatically assisted grader machine

GNSS APPLICATIONS AND EQUIPMENT

> "Farming looks mighty easy when your plough is a pencil, and you're a thousand miles from the corn field."
> —Dwight D. Eisenhower, 34th U.S. president.

Figure 63 Precisely aligned crop rows planted by GNSS-guided machinery

Harvesting strawberries

The agriculture industry is facing a shortage of workers. In a recent survey, more than 40% of farmers stated they have been unable to obtain the labour they need.

Motivated by the threat of a dwindling workforce, Harvest CROO Robotics created a fully autonomous vehicle to harvest strawberries, a traditionally labour-intensive job. The vehicle uses innovative technology to determine the ripeness of the fruit and robot pickers to harvest the ripe fruit.

The autonomous vehicle is equipped with two GNSS receivers using PPP. The repeatable pass-to-pass accuracy of the GNSS receivers with PPP allows the vehicle to reliably navigate the rows of strawberries without damaging the crops.

An article (Field of Dreams) about how GNSS has helped with harvesting strawberries is in the 2019 Velocity magazine available at: resources.hexagonpositioning.com/field-of-dreams.

Autonomous crop spraying

Crop spraying is used to protect crops and increase productivity. Methods typically used for spraying crops, such as backpacks, tractors and manned aircraft, have limitations and risks. Backpack spraying can lead to high levels of chemical exposure and associated health risks to the operator. Tractors can crush crops as they spray, hurting the farmer's profitability. Manned aircraft are limited by the availability of airport facilities and can be expensive, making it especially difficult for smaller farms to use them.

Using Unmanned Aircraft Systems (UAS) to spray crops reduces risks to operators, crop damage and the labour required to complete the task. A dual-antenna GNSS receiver with RTK positioning technology provides the precise positioning and heading direction

GNSS APPLICATIONS AND EQUIPMENT

Figure 64 GNSS-based surveying equip-

information needed to make crop spraying successful.

An article (Farming Smarter) about how GNSS provides the precise positioning required for UAS spraying is in the 2018 Velocity magazine available at: resources.hexagonpositioning.com/farming-smarter.

Construction

GNSS information can be used to position the cutting edge of a blade (on a bulldozer or grader, for example) or a bucket (excavator), and to compare this position against a 3D digital design to compute cut/fill amounts. "Indicate systems" provide the operator with visual cut/fill information via a display or light bar, and the operator manually moves the machine's blade or bucket to get to grade. Automatic systems for bulldozers/graders use the cut/fill information to drive the hydraulic controls of the machine to automatically move the machine's blade to grade. The use of 3D machine control dramatically reduces the number of survey stakes required on a job site, reducing time and costs. Productivity studies have repeatedly shown that the use of 3D machine control results in work being completed faster, more accurately and with significantly less rework than conventional construction methods.

Surface mining

GNSS information is being used to efficiently manage the mining of an ore body and the movement of waste material. GNSS equipment installed on shovels and haul trucks provides position information to a computer-controlled dispatch system to optimally route haul trucks to and from each shovel. Position information is also used to track each bucket of material extracted by the shovel to ensure that it is routed to the appropriate location in the mine (crusher, waste dump, leach pad). Position information is used by blast hole drills to improve fracturing of the rock material and control the depth of each hole that is drilled to keep the benches level. Multi-constellation GNSS is particularly advantageous in a surface mining environment due to the obstructions caused by the mine's walls: more satellites means more signal availability.

Not only is GNSS positioning being used to optimize operations within the mine, but it is also enabling automated road trains for safer, more efficient mine-to-port transport.

GNSS APPLICATIONS AND EQUIPMENT

An article (Perception, Positioning and a Long Haul Approach to Autonomy) about how GNSS is used for autonomous platooning in an Australian iron ore mine is in the 2022 Velocity magazine available at: resources.hexagonpositioning.com/perception-positioning-long-haul.

Automated blast hole drilling

Automated drills are used in surface mines to increase safety and productivity. A single operator safely located in the control room can operate and monitor up to five automated drills.

The blast holes drilled by the automated drills must be very precise, both horizontally and vertically. The position of the holes (horizontal accuracy) is critical in controlling rock fragmentation. Rock fragments that are too large or too fine can increase wear on the rock crushers used to process the material. Hole depth (vertical accuracy) is important for creating a flat bench.

Three GNSS technologies are used on the automated drills: RTK, heading and multi-constellation. RTK provides the precise positioning needed to accurately locate the blast holes. Heading provides the alignment of the drill to ensure the holes are drilled perpendicular. Multi-constellation receivers compensate for signal blockages common in the high wall environment typical of surface mines.

An article (The Pit, The Bit and The Benefit) about how GNSS is used in automated drilling is in the 2013 Velocity magazine available at: resources.hexagonpositioning.com/pit-bit-benefit.

Survey

GNSS-based surveying reduces the amount of equipment and labour required to determine the position of points on the surface of the Earth compared with previous surveying techniques. Using GNSS, it is possible for a single surveyor to accomplish in one day what might have taken a survey crew of three people a week to complete.

Determining a new survey position once required measuring distances and bearings from an existing (known) survey point to the new point. This required using a theodolite to measure angular differences and metal "chains" (long heavy tape measures) pulled taught to minimise sag and accurately measure distances. If the new and existing survey points were separated by a large distance, the process would involve multiple setups of the theodolite, then multiple angular and distance measurements.

Using GNSS, surveyors can now set up a DGNSS or RTK base station over an existing survey point and a DGNSS or RTK rover over the new point, then record the position measurement at the rover. This simplification shows why the surveying industry was one of the early civilian adopters of GNSS technology.

Seismic survey sensors

In a seismic survey, sound waves are sent from a source (explosives or a thumper truck) through the ground to an array of sensors (geophones). Knowing the exact location and orientation of the geophones is critical for a successful survey.

Placement of traditional geophones is a two-step process. First, a team surveys the area and places markers for each geophone. Later, a second team places the geophones precisely on the marked locations and then orients them using a direction measuring device, such as a compass.

Using GNSS-enabled geophones

An Introduction to GNSS, Third Edition   89

GNSS APPLICATIONS AND EQUIPMENT

eliminates the need to survey the area first. The GNSS-enabled geophones have a GNSS receiver and dual-antennas integrated into the geophone. The receiver and dual-antennas allow the geophone to not only determine its exact location, but also its orientation.

An article (Building a Better Geophone) about how GNSS is used to simplify geophone placement is in the 2013 Velocity magazine available at: resources.hexagonpositioning.com/building-better-geophone.

Aerial photogrammetry

Aerial photogrammetry refers to the recording of images of the ground (photographs, for example) from an elevated position, such as an aircraft. Systems of this type are now more generally referred to as "remote sensing" since the images can be taken from aircraft or from satellites.

In the past, images would have to be manually corrected for orientation, perspective, height and location of the camera and manually "stitched" together. This manual process would be based on the accurate alignment of known points in adjacent pictures.

By integrating the camera with GNSS+INS, it is now possible to automate the process in real-time or post-mission to "transfer" the location accuracy of the aircraft determined from GNSS to the image.

Aerial photographs are used in online map systems such as Google Earth. Many of us have found our houses, and perhaps even our cars, through these applications.

GNSS technology has also been integrated with LiDAR (Light Detection and Ranging), an optical remote sensing technology used to measure the range to distant targets. It is possible to detect and image a feature or object down to the wavelength, which at LiDAR frequencies is less than a millionth of a metre.

Mapping wildfires

To battle wildfires, firefighters need to know the locations of the fires and any hotspots. Using an airplane equipped with an infrared imaging sensor and a GNSS+INS system, locations of fires and hotspots can be projected on topographical or 3D terrain maps.

An article (Custom Airborne Mapping Solutions) about how GNSS is used in aerial mapping is in the 2014 Velocity magazine available at: resources.hexagonpositioning.com/custom-airborne-mapping-solutions.

Ground mapping

Products have been developed that take 360-degree panoramic photographs to support the presentation of geometrically correct images on a computer screen. These images are continuous and precisely positioned. GNSS and IMU data are recorded before the panoramic photographs are taken. Position and attitude data is programmed

Figure 65 Aerial image of Niagara Falls

GNSS APPLICATIONS AND EQUIPMENT

> "The clock, not the steam engine, is the key-machine of the modern industrial age."
>
> —**Lewis Mumford**, American historian of technology and science.

into the cameras, allowing onscreen determination of positions of objects in the photos or measurements between objects.

Infrastructure visualisation

Using a LiDAR combined with a GNSS+INS, a user can capture comprehensive visual information of key infrastructure, such as oil and gas pipelines. This visual information provides the state, location and positioning of the infrastructure and its related assets. It also assists in planning for maintenance and modifications.

An article (Seeing is Believing) about how GNSS and LiDAR are used for infrastructure visualisation is in the 2014 Velocity magazine available at: resources.hexagonpositioning.com/seeing-is-believing.

Geospatial information systems (GIS)

A GIS captures, stores, analyses, manages and presents data that is linked to location. The data may consist of, for example, environmental or resource data. GIS is also used to map attributes for insurance companies, municipal planning, utility companies and others. The positions associated with the data can be provided from a GNSS receiver. GIS applications can generate detailed contour maps from the data and present these maps in a digital form, as illustrated in **Figure 66**.

Improving weather predictions

Weather forecasting impacts our society and economy in very tangible ways, influencing important decisions business leaders and public safety officials make every day. One of the important components in accurate forecasting is knowing the water vapour content in the atmosphere. Water vapour plays a critical role in the global hydrologic system, and is a key component in cloud formation and the lower atmosphere's chemistry. Despite its important role in forecasting, water vapour is difficult to observe, especially during severe weather.

GNSS receivers are being used to help observe water vapour, one of the major components in the tropospheric delay of GNSS signals. These signal delays can be estimated and used to retrieve the total column water vapour, also known as Integrated Precipitable Water Vapour (IPW).

A network of GNSS receivers spread across the forecasting area makes it possible to observe IPW accurately and reliably in all weather conditions. These observations can be fed into weather

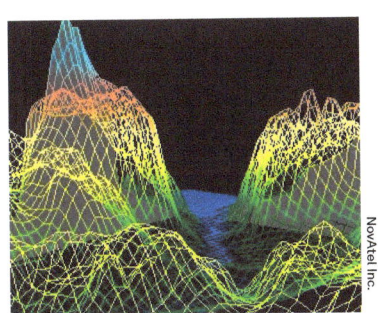

Figure 66 GIS data output

forecast models to better predict impactful weather events such as hurricanes and large storms.

An article (Using GPS for Better Weather Prediction) about using GPS receivers to measure water vapour is in the 2018 Velocity magazine available at:resources.hexagonpositioning.com/using-gps-weather-prediction.

Time applications

As we mentioned in earlier chapters, time accuracy is critical for GNSS position determination. This is why GNSS satellites are equipped with atomic clocks that are accurate to nanoseconds. As part of the position determining process, the local time of GNSS receivers becomes synchronised with the very accurate satellite time.

This time information, by itself, has many applications, including the synchronisation of communication systems, electrical power grids and financial networks. GNSS-derived time

Figure 67 Marine application of GNSS

works well for any application where precise timing is needed by devices that are dispersed over a wide area.

Seismic monitors that are synchronised with GNSS satellite clocks can be used to determine the epicentre of an earthquake by triangulation based on the exact time the earthquake was detected by each monitor.

Marine applications

In addition to dramatically improving marine navigation, GNSS is also being applied to a broad range of marine applications, such as oil rig positioning, underwater cable and pipeline installation and inspection, rescue and recovery and the dredging of ports and waterways.

GNSS-equipped sonobuoys

An interesting application of GNSS is the use of GNSS-equipped sonobuoys in underwater sonar systems.

Sonobuoys are dropped from aircraft over an area of interest but are left to float autonomously. The sonobuoys detect approaching ships and other hazards in the water by transmitting sound waves through the water, detecting reflections from vessels and objects, and determining the time it takes for the "echo" to be received. The data comes up to the sonobuoy's float and then is transmitted with GNSS positioning data, over a radio link to a survey ship. The survey ship collects and analyses sonar data from a large number of sonobuoys, then determines and displays the location of ships and objects in the area of interest.

Seafloor mapping

Knowing the depth of the seafloor in ports and navigation channels is critical to safe marine navigation. Maps of the ports and channels are created using bathymetric sonar systems.

Bathymetric sonar systems mounted in a marine vessel bounce sound waves off the seafloor to determine the depth of the water. Using these depth

GNSS APPLICATIONS AND EQUIPMENT

measurements, a map of the seafloor is created. For the seafloor maps to be accurate, the exact location of the vessel on the surface of the water must be known. A GNSS+INS integrated with the sonar system provides the precise location of the vessel for each sonar measurement. The GNSS+INS also provides the vertical location of the vessel to compensate for waves.

An article (Sounding the Depths) about how GNSS works with bathymetric sonar is in the 2014 Velocity magazine available at: resources.hexagonpositioning.com/sounding-the-depths.

Advancements in autonomous marine vessels are enabling unmanned surveying and other critical offshore operations either via remote operations from shore or from an office anywhere in the world. An article (Safe Autonomy at Sea) about how GNSS equipment is applied to marine autonomy is in the 2022 Velocity magazine available at: resources.hexagonpositioning.com/safe-autonomy-at-sea.

Unmanned vehicles

An unmanned vehicle is unoccupied but under human control, whether radio-controlled or automatically guided by a GNSS-based application. There are many types of unmanned vehicles, including: Unmanned Ground Vehicle (UGV), Unmanned Aerial Vehicle (UAV), Unmanned Surface Vehicle (USV) and Unmanned Underwater Vehicle (UUV).

Initially, unmanned vehicles were used primarily by the defence industry. However, as the unmanned vehicle market has grown and diversified, their commercial use has grown. Some of the current civilian uses for unmanned vehicles are: search and rescue, crop monitoring, wildlife conservation, aerial

Figure 68 Reaper Unmanned Aerial Vehicle (UAV)

photography, environmental research, infrastructure inspection, bathymetry, landmine detection and disposal, HAZMAT inspection and disaster management. As the civilian unmanned vehicle market expands, so will their civilian use.

Hurricane research

Knowing where a hurricane will make landfall and how powerful it will be are important to properly prepare for the storm. While meteorologists are good at predicting the potential path of a hurricane, it is much harder to predict how powerful the storm will be when it arrives.

To learn more about what causes a hurricane to rapidly increase or decrease in intensity, NASA is using two long-range UAVs to study storms while they are still far out over the Ocean. Onboard the UAVs are meteorological instruments that monitor the environmental conditions in and around the storm. The UAVs also have a GNSS+INS that records the UAV location and attitude for each measurement taken by the meteorological instruments. An accurate

GNSS APPLICATIONS AND EQUIPMENT

UAV location and attitude is necessary for the measurement taken to be useful.

An article (Joining the Hunt) about the NASA project to study hurricane intensification is in the 2014 Velocity magazine available at: resources.hexagonpositioning.com/joining-the-hunt.

Orion spacecraft parachute testing

Before the Orion spacecraft can be used for manned space missions, NASA must know that Orion can safely land on Earth. A key aspect of returning the astronauts safely to Earth is slowing the Orion spacecraft from its incredibly high re-entry speed of close to 32,000 km/h (19,884 mph) to less than 36 km/h (22 mph). This is the job of the Capsule Parachute Assembly System (CPAS).

To test the CPAS, NASA created two unmanned test vehicles. These test vehicles were dropped out of a C-17 aircraft from altitudes as high as 10,668 metres (35,000 feet). A GNSS+INS installed in the test vehicles measured the vertical velocity to test the parachute system's effectiveness.

An article (Put to the Test) about the Orion CPAS testing is in the 2014 Velocity magazine available at: resources.hexagonpositioning.com/put-to-the-test.

Landing an unmanned helicopter on a ship

The autonomous landing of an unmanned helicopter is already challenging as the navigation system needs to deal with movement of the helicopter caused by wind. This challenge is greatly increased when trying to land on a ship at sea. Not only is the helicopter's position changing based on its movement and the effects of the wind, but the ship is also moving independently based on its movement and the effects of both the wind and the sea.

When landing a helicopter on a ship, the relative distance between the helicopter landing gear and the flight deck of the ship is much more important than the absolute position of the helicopter and ship. GNSS+INS technology installed on both the helicopter and the ship are used to determine this relative distance. The GNSS+INS on the ship calculates its position and sends that information to the GNSS+INS on the helicopter. The GNSS+INS on the helicopter uses the position sent from the ship along with its own position to calculate the relative distance and direction between the ship and helicopter. Using this relative distance and direction, the unmanned helicopter is able to autonomously approach and land on the ship's flight deck.

An article (From Fledgling to Flight) about the landing of the unmanned Little Bird helicopter on a moving ship is in the 2013 Velocity magazine available at: resources.hexagonpositioning.com/from-fledgling-to-flight.

Delivering critical medical supplies

Delivering essential medical supplies to hospitals and medical clinics in remote areas is often challenging. Long distances and poor infrastructure can delay deliveries of supplies that are desperately needed to treat patients. Using UAVs can speed the delivery of these critical medical supplies.

An article (Zipline: Deploying Drones to Save Lives) about delivering medical supplies to remote hospitals and medical clinics is in the 2018 Velocity magazine available at: resources.hexagonpositioning.com/deploying-drones-to-save-lives.

GNSS APPLICATIONS AND EQUIPMENT

Defence

The defence sector makes broad use of GNSS technology, including:

NAVIGATION:
Using GNSS receivers, soldiers and pilots can navigate unfamiliar terrain or conduct night-time operations. Most foot soldiers now carry hand-held GNSS receivers.

SEARCH AND RESCUE:
If a plane crashes and has a search and rescue beacon equipped with a GNSS receiver, it can be located more quickly.

RECONNAISSANCE AND MAP CREATION:
The military uses GNSS to create maps of uncharted or enemy territory. They can also mark reconnaissance points using GNSS.

UNMANNED VEHICLES:
Unmanned vehicles are used extensively in military applications, including reconnaissance, logistics, target and decoy, mine detection, search and rescue, research and development and missions in unsecured or contaminated areas. An article (CAMCOPTER® S-100 UAS) about how anti-jamming technology helps unmanned vehicles complete missions in GNSS denied environments is in the 2016 Velocity magazine at: resources.hexagonpositioning.com/camcopter-s-100-uas.

MUNITIONS GUIDANCE:
Precision munitions use GNSS to ensure they land on target.
An article (GAJT Put to the Test) about the use of anti-jamming technology with guided munitions is in the 2017 Velocity magazine available at: resources.hexagonpositioning.com/GAJT-put-to-the-test.

GNSS equipment

The first generation of commercial GNSS receivers cost well over $100,000. Now, GNSS receivers are built into smartphones. Vendors have developed a wide array of equipment to support the incredible range of possible GNSS applications. As illustrated in **Figure 69**, GNSS equipment consists of receivers, antennas and supporting software at varying levels of integration and performance.

Depending on the application, the antenna and receiver might be separate entities, or they may be integrated into a single package, as in a hand-held GNSS receiver. GNSS equipment may be further integrated with application equipment such as survey or hydrographic instrumentation or a transport vessel.

GNSS equipment specifications and features depend on the application. To illustrate, users need to consider the following when selecting GNSS equipment for a particular use:

ACCURACY:
Applications such as survey may require centimetre-level (inch-level) accuracy. Others, such as positioning for hiking, may only require accuracy to within tens of metres (tens of yards). Some applications require absolute accuracy; that is, position defined accurately relative to an actual reference point or location. Others may require accuracy relative to a previous position. If high-precision accuracy is obtained through the application of differential GNSS, it may be desirable for the differential service to be integrated in the same package as the GNSS receiver, for example, the radio link to the base station or rovers.

An Introduction to GNSS, Third Edition

GNSS APPLICATIONS AND EQUIPMENT

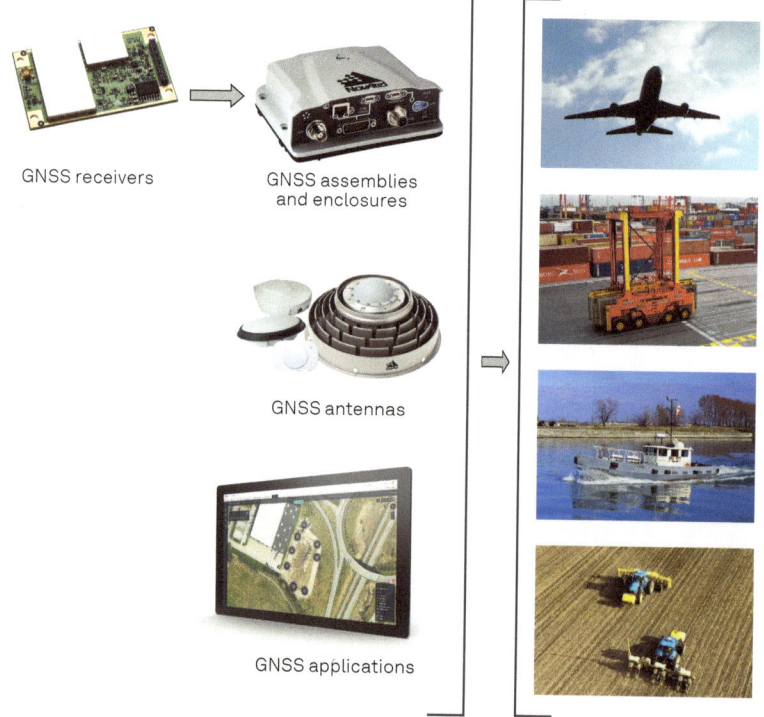

Figure 69 Examples of GNSS equipment

ACQUISITION TIME:
For some applications, users may require a fast "time to first fix," the time required by a GNSS receiver to achieve a position solution. For other applications, it may not be important that the "fix" be available quickly. The trade-off to achieving fast acquisition time is increased probability of a wrong position/fix.

RELIABILITY:
Reliability addresses the question, "Do you need the answer (position and time) to be correct every time?"

AVAILABILITY:
Equipment may be required to provide positioning service continuously, even in areas where signals from satellites are blocked. As we have discussed, these applications may best be served by integrating GNSS and INS equipment.

Equipment may need to support multiple constellations and frequencies, and perform well in environments characterised by a high level of multipath interference. Remember from Chapter 4, multipath interference occurs because some of the signal energy transmitted by the satellite is reflected (and therefore delayed) on the way to the receiver. In the selection of GNSS equipment, there will almost always be trade-offs between accuracy, acquisition time, reliability and availability.

ENVIRONMENTAL:
User equipment may have to operate

over wide temperature and humidity ranges, at high altitudes or in dusty environments. The equipment may need to be waterproof to rain or submersion.

SHOCK AND VIBRATION:
Equipment may be subjected to high levels of shock and vibration, such as that characteristic of industrial vehicles.

PORTABILITY:
Depending on the application, the equipment may need to be portable, such as a hand-held device for hiking or survey.

REGULATORY:
Regulatory compliance will vary with the jurisdiction in which the user is operating, for example:

- Emissions standards, such as FCC Part 15
- Compliance with the European Union's Restriction of Hazardous Substances (RoHS) directive
- WEEE, the European Community directive that imposes responsibility for the disposal of waste electrical and electronic equipment on the equipment manufacturer

DATA STORAGE:
Receivers may be required to store time-stamped range or position information for applications that will use this information post-mission.

PHYSICAL SIZE AND POWER CONSUMPTION:
The user may require a receiver or antenna with a small form-factor and low power consumption for size-limited applications, such as UAVs.

USER INTERFACE:
The manner through which the user interacts with the equipment is important; for example, a keypad for entering commands, a screen for viewing position data on a map or connectors for outputting data to other devices.

COMPUTATIONAL REQUIREMENTS:
Users may require that the equipment provide computed data such as velocity or heading.

COMMUNICATIONS:
Position may only be useful if it is communicated to another device over, for example, a cellular radio link.

FUTURE-PROOF:
Although some GNSS signals and constellations may not yet be available, users may require some assurance that they will be able to use these signals and constellations once they are available.

Closing remarks

Like "cyberspace," GNSS is already here, and its broad acceptance and application is based on a track record of exceptional performance and reliability. Industry and government agencies are continually enhancing technology and infrastructure to enable the development of new GNSS-based solutions.

In this chapter, we have provided a broad sample of current GNSS applications to illustrate just how beneficial GNSS is both in terms of cost efficiency and safety of life applications. GNSS technology is becoming truly ubiquitous — a prevalent, taken-for-granted technology in almost everything we do. GNSS anywhere and anytime is here.

"I used to think that cyberspace was fifty years away. What I thought was fifty years away, was only ten years away. And what I thought was ten years away… it was already here. I just wasn't aware of it yet."

–Bruce Sterling, American science fiction author.

APPENDIX A – ACRONYMS

Appendices

This section includes the following appendices, which include general or supplementary information about GNSS:

- **Appendix A**–Acronyms
- **Appendix B**–GNSS Glossary
- **Appendix C**–Standards and References
- **Appendix D**–Acknowledgements

Acronym	Definition
1PPS	One Pulse Per Second
ADR	Accumulated Doppler Range
AFSCN	Air Force Satellite Control Network
AltBOC	Alternate Binary Offset Carrier
AMSAT	American Satellite
ARNS	Aeronautical Radio Navigation Services
ARP	Antenna Reference Point
AVL	Automatic Vehicle Location
BDS	BeiDou Navigation Satellite System
BOC	Binary Offset Carrier
C/A Code	Coarse/Acquisition Code
CASM	Coherent Adaptive Subcarrier Modulation
CD	Clock Drift
CDGPS	Canada-Wide Differential GPS
CDMA	Code Division Multiple Access
CE	Conformité Européenne (also known as CE Mark)
CMG	Course Made Good
CNAV	Civil Navigation
C/No	Post Correlation Carrier to Noise Ratio in dB-Hz
COG	Course Over Ground
COGO	Coordinate Geometry
COSPAS	Cosmitscheskaja Sistema Poiska Awarinitsch Sudow (Russian: space system for search of vessels in distress)
CRPA	Controlled Reception Pattern Antenna
CS	Commercial Service
CTP	Conventional Terrestrial Pole
CTS	Conventional Terrestrial System
dB	Decibel
dBm	Decibel Relative to 1 milliWatt
DGNSS	Differential Global Navigation Satellite System
DGPS	Differential Global Positioning System
DOP	Dilution Of Precision
DR	Dead Reckoning
e	Eccentricity
EC	European Commission
ECEF	Earth-Centred-Earth-Fixed
EGNOS	European Geostationary Navigation Overlay System
ESA	European Space Agency
FAA	Federal Aviation Administration
FCC	Federal Communication Commission
FDMA	Frequency Division Multiple Access
FKP	Flächen Korrectur Parameter (Plane Correction Parameter) German
FOC	Full Operational Capability
FOG	Fiber Optic Gyro
GAGAN	GNSS Aided GEO Augmented Navigation (India)
GBAS	Ground Based Augmentation System
GCC	Galileo Control Centre
GDOP	Geometric Dilution Of Precision
GEO	Geostationary Earth Orbit
GIC	GNSS Integrity Channel
GIS	Geospatial Information System
GLONASS	Globalnaya Navigazionnaya Sputnikovaya Sistema (Russian Global Navigation Satellite System)
GMS	Ground Mission Segment
GMT	Greenwich Mean Time
GNSS	Global Navigation Satellite System
GPS	Global Positioning System
GRAS	Ground-based Regional Augmentation System (Australia)
GRC	Galileo Reception Chain
GRCN	Galileo Reception Chain Non-PRS
GSS	Galileo Sensor Stations
GSTB	Galileo System Test Bed
GTR	Galileo Test Receiver
GTS	Galileo Test Signal Generator
GUS	Ground Uplink Station
GUST	WAAS GUS-Type 1
GUSTR	WAAS GUST Type-1 Receiver
HDOP	Horizontal Dilution Of Precision
hex	Hexadecimal
HTDOP	Horizontal Position and Time Dilution Of Precision
Hz	Hertz
I and Q	In-Phase and Quadrature (Channels)
I Channel	In-Phase Data Channel
ICP	Integrated Carrier Phase
IEC	International Electrotechnical Commission

APPENDIX A – ACRONYMS

IERS	International Earth Rotation Service
IGP	Ionospheric Grid Point
IGRF	International Geometric Reference Field
IGS CB	International GNSS Service Central Bureau
IGSO	Inclined Geosynchronous Orbit
IMLA	Integrated Multipath Limiting Antenna
IMU	Inertial Measurementt Unit
INS	Inertial Navigation System
I/O	Input/Output
IODE	Issue of Data (Ephemeris)
IOV	In-Orbit Validation
IRNSS	Indian Regional Navigation Satellite System
ITRF	International Terrestrial Reference System
KASS	Korean Augmentation Satellite System
L1	The 1575.42 MHz GPS carrier frequency including C/A and P-Code
L1C	GPS L1 civilian frequency
L2	The 1227.60 MHz 2nd GPS carrier frequency (P-Code only)
L2C	The L2 civilian code transmitted at the L2 frequency (1227.6 MHz)
L5	The 1176.45 MHz 3rd civil GPS frequency that tracks carrier at low signal-to-noise ratios
LAAS	Local Area Augmentation System
LiDAR	Light Detection and Ranging
LGF	LAAS Ground Facility
LNA	Low Noise Amplifier
LORAN	LOng RANge Navigation System
MAT	Multipath Assessment Tool
mBOC	Multiplexed Binary Offset Carrier
MEDLL	Multipath Estimating Delay Lock Loop
MEO	Medium Earth Orbit
MHz	MegaHertz
ms	Millisecond
MSAS	MTSAT Satellite Based Augmentation System (Japan)
MSAT	Mobile Satellite
MSL	Mean Sea Level
MTSAT	Multi-Functional Transport Satellite
NASA	National Aeronautics and Space Administration (U.S.)
NavIC	Navigation Indian Constellation
NAVSTAR	NAVigation Satellite Timing And Ranging (synonymous with GPS)
NMEA	National Marine Electronics Association
ns	Nanosecond
OEM	Original Equipment Manufacturer
OS	Open Service
PAC	Pulsed Aperture Correlator
PCO	Phase Centre Offset
P-Code	Precise Code
PDOP	Position Dilution Of Precision
PDP	Pseudorange/Delta-Phase
PE-90	Parameters of the Earth 1990 (see PZ-90)
PIN	Position Indicator
PLL	Phase Lock Loop
PNT	Positioning, Navigation and Timing
PPM	Parts Per Million
PPP	Precise Point Positioning
PPS	Precise Positioning Service
PPS	Pulse Per Second
PRN	Pseudorandom Noise
PRS	Public Regulated Service
PSR	Pseudorange
PVT	Position Velocity Time
PZ-90	Parametry Zemli 1990 (see PE-90)
Q Channel	Quadrature Data-Free Channel
QSO	Quasi-Zenith Orbit
QZSS	Quasi-Zenith Satellite System
RCC	Rescue Coordination Centre
RF	Radio Frequency
RINEX	Receiver Independent Exchange Format
RLG	Ring Laser Gyro
RoHS	Restriction of the use of Hazardous Substances
RMS	Root Mean Square
RPDP	Relative PDP
RS	Restricted Service
RSS	Residual Solution Status
RTCA	Radio Technical Commission for Aeronautics
RTCM	Radio Technical Commission for Maritime Services
RTK	Real-Time Kinematic
SA	Selective Availability
SAR	Search and Rescue
SARSAT	Search and Rescue Satellite Aided Tracking
SBAS	Satellite Based Augmentation System
SD	Standard Deviation
SDCM	System for Differential Corrections and Monitoring
SG	Signal Generator
SGS-90	Soviet Geodetic System 1990
SI	Système Internationale
SiS	Signal in Space
SNAS	Satellite Navigation Augmentation System (China)
SNR	Signal-to-Noise Ratio
SOL	Safety of Life
SPS	Standard Position Service
SPAN	Synchronised Position Attitude Navigation
SV	Space Vehicle
SVID	Space Vehicle Identifier
SVN	Space Vehicle Number
TDOP	Time Dilution Of Precision
TTFF	Time-To-First-Fix
TTNL	Time to Narrow Lane
UHF	Ultra High Frequency
USGS	United States Geological Survey
UAV	Unmanned Aerial Vehicle
UGV	Unmanned Ground Vehicle
UNB	University of New Brunswick
USV	Unmanned Surface Vehicle

APPENDIX B – GNSS GLOSSARY

Absolute Accuracy
In GNSS positioning, absolute accuracy is the degree to which the position of an object on a map conforms to its correct location on the Earth, according to an accepted coordinate system.

Acquisition
The process of locking onto a satellite's C/A code and P-code. A receiver acquires all available satellites when it is first powered up, then acquires additional satellites as they become available and continues tracking them until they become unavailable.

Accumulated Doppler Range (ADR)
Carrier phase, in cycles. [See *Carrier Phase Measurements*].

Almanac
A set of orbit parameters that allows calculation of approximate GNSS satellite positions and velocities. The almanac is used by a GNSS receiver to determine satellite visibility and as an aid during acquisition of GNSS satellite signals.

Almanac Data
A set of data which is downloaded from each satellite over the course of 12.5 minutes. It contains orbital parameter approximations for all satellites, GNSS to Universal Standard Time (UTC) conversion parameters, and single-frequency ionospheric model parameters.

Ambiguity
The integer number of carrier cycles between a satellite and receiver.

Anti-Spoofing
Denial of the P-code by the Control Segment is called Anti-Spoofing. It is normally replaced by encrypted Y-code. [See *P-Code* and *Y-Code*]

Antipodal Satellites
Antipodal satellites are satellites in the same orbit plane separated by 180 degrees in argument of latitude.

Baseline
The line between a pair of stations for which simultaneous GNSS data has been collected.

Base Station
The GNSS receiver which is acting as the stationary reference. It has a known position and transmits messages for the rover receiver to use to calculate its position.

Bearing
The horizontal direction of one terrestrial point from another terrestrial point, expressed as the angular distance from a reference direction, usually measured from 000° at the reference direction clockwise through 360°. The reference point may be true north, magnetic north or relative (vehicle heading).

BeiDou Navigation System (BDS)
BeiDou is China's global navigation satellite system.

Broadcast Ephemerides
A set of parameters which describes the location of satellites with respect to time, and which is transmitted (broadcast) from the satellites.

Carrier
The steady transmitted RF signal whose amplitude, frequency or phase may be modulated to carry information.

Carrier Phase Ambiguity
The number of integer carrier phase cycles between the user and the satellite at the start of tracking (sometimes ambiguity for short).

Carrier Phase Measurements
These are "Accumulated Doppler Range" (ADR) measurements. They contain the instantaneous phase of the signal (modulo 1 cycle) plus some arbitrary number of integer cycles. Once the receiver is tracking the satellite, the integer number of cycles correctly accumulates the change in range seen by the receiver. When a "lock break" occurs, this accumulated value can jump an arbitrary integer number of cycles (this is called a cycle slip).

C-Band
C-Band is the original frequency allocation for communications satellites. C-Band uses 3.7-4.2 GHz for downlink and 5.925-6.425 GHz for uplink.

Coarse Acquisition (C/A) Code
A pseudorandom string of bits that is used primarily by commercial GNSS receivers to determine the range to the transmitting GNSS satellite. The 1023 chip GPS C/A code repeats every 1 ms giving a code chip length of 300 metres (328 yards), which is very easy to lock onto.

APPENDIX B – GNSS GLOSSARY

Control Segment
The Master Control Station and the globally dispersed Reference Stations used to manage the GNSS satellites, determine their precise orbital parameters and synchronise their clocks.

Coordinated Universal Time (UTC)
This time system uses the second-defined true angular rotation of the Earth measured as if the Earth rotated about its Conventional Terrestrial Pole. However, UTC is adjusted only in increments of one second. The time zone of UTC is that of Greenwich Mean Time (GMT).

Course
The horizontal direction in which a vessel is to be steered or is being steered; the direction of travel through the air or water. Expressed as angular distance from reference north (either true, magnetic, compass or grid), usually 000° (north), clockwise through 360°. Strictly, the term applies to direction through the air or water, not the direction intended to be made good over the ground [See *Track Made Good*]. Differs from heading.

Course Made Good (CMG)
The single resultant direction from a given point of departure to a subsequent position; the direction of the net movement from one point to the other. This often varies from the track caused by inaccuracies in steering, currents, cross-winds, etc. This term is often considered to be synonymous with Track Made Good, however, Course Made Good is the more correct term.

Course Over Ground (COG)
The actual path of a vessel with respect to the Earth (a misnomer in that courses are directions steered or intended to be steered through the water with respect to a reference meridian); this will not be a straight line if the vessel's heading yaws back and forth across the course.

Cycle Slip
When the carrier phase measurement jumps by an arbitrary number of integer cycles. It is generally caused by a break in the signal tracking due to shading or some similar occurrence.

Dead Reckoning (DR)
The process of determining a vessel's approximate position by applying DR from its last known position a vector or a series of consecutive vectors representing the run that has since been made, using only the courses being steered, and the distance run as determined by log, engine rpm or calculations from speed measurements.

Destination
The immediate geographic point of interest to which a vessel is navigating. It may be the next waypoint along a route of waypoints or the final destination of a voyage.

Differential GNSS (DGNSS)
A technique to improve GNSS accuracy that uses pseudorange errors, at a known location, to improve the measurements made by other GNSS receivers within the same general geographic area.

Dilution of Precision (DOP)
A numerical value expressing the confidence factor of the position solution based on current satellite geometry. The lower the value, the greater the confidence in the solution. DOP can be expressed in the following forms.
- GDOP: Uncertainty of all parameters (latitude, longitude, height, clock offset)
- PDOP: Uncertainty of 3D parameters (latitude, longitude, height)
- HTDOP: Uncertainty of 2D and time parameters (latitude, longitude, time)
- HDOP: Uncertainty of 2D parameters (latitude, longitude)
- VDOP: Uncertainty of height parameter
- TDOP: Uncertainty of clock offset parameter

Doppler
The change in frequency of sound, light or other wave caused by movement of its source relative to the observer.
- **Theoretical Doppler:** The expected Doppler frequency based on a satellite's motion relative to the receiver. It is computed using the satellite's coordinates and velocity, and the receiver's coordinates and velocity.
- **Apparent Doppler:** Same as Theoretical Doppler of satellite above, with clock drift correction added.
- **Instantaneous Carrier:** The Doppler frequency measured at the receiver, at that epoch.

APPENDIX B – GNSS GLOSSARY

Doppler Aiding
A signal processing strategy, which uses a measured Doppler shift to help a receiver smoothly track the GNSS signal, to allow more precise velocity and position measurement.

Double-Difference
A mathematical technique comparing observations by differencing between receiver channels and then between the base and rover receivers.

Double-Difference Carrier Phase Ambiguity
Carrier phase ambiguities which are differenced between receiver channels and between the base and rover receivers. They are estimated when a double-difference mechanism is used for carrier phase positioning (sometimes double-difference ambiguity or ambiguity, for short).

Earth-Centered-Earth-Fixed (ECEF)
This is a coordinate system which has the X-axis in the Earth's equatorial plane pointing to the Greenwich prime meridian, the Z-axis pointing to the north pole, and the Y-axis in the equatorial plane 90° from the X-axis with an orientation which forms a right-handed XYZ system.

Eccentricity (e)
A dimensionless measurement defined for a conic section where e=0 is a circle, 0<e<1 is an ellipse, e=1 is a parabola and e>1 is a hyperbola. For an ellipse, larger values of e correspond to a more elongated shape. The eccentricity of GNSS satellite orbit is typically .02.

Elevation
The angle from the horizon to the observed position of a satellite.

Ellipsoid
A smooth mathematical surface which represents the Earth's shape and very closely approximates the geoid. It is used as a reference surface for geodetic surveys.

Ellipsoidal Height
Height above a defined ellipsoid approximating the surface of the Earth.

Ephemeris
A set of satellite orbit parameters that are used by a GNSS receiver to calculate precise GNSS satellite positions and velocities. The ephemeris is used in the determination of the navigation solution and is updated periodically by the satellite to maintain the accuracy of GNSS receivers.

Ephemeris Data
The data downlinked by a GNSS satellite describing its own orbital position with respect to time.

Epoch
Strictly a specific point in time. Typically when an observation is made.

Fixed Ambiguity Estimates
Carrier phase ambiguity estimates which are set to a given number and held constant. Usually they are set to integers or values derived from linear combinations of integers.

Fixed Discrete Ambiguity Estimates
Carrier phase ambiguities which are set to values that are members of a predetermined set of discrete possibilities, and then held constant.

Fixed Integer Ambiguity Estimates
Carrier phase ambiguities which are set to integer values and then held constant.

Galileo
The European Union's global navigation satellite system.

Geometric Dilution of Precision (GDOP)
[See *Dilution of Precision (DOP)*]

Geoid
The shape of the Earth if it were considered as a sea level surface extended continuously through the continents. The geoid is an equipotential surface coincident with mean sea level to which at every point the plumb line (direction in which gravity acts) is perpendicular. The geoid, affected by local gravity disturbances, has an irregular shape.

Geodetic Datum
The reference ellipsoid surface that defines the coordinate system.

Geostationary
A satellite orbit along the equator that results in a constant, fixed position over a particular reference point on the Earth's surface.

Geosynchronous
A satellite orbit with an orbital period matching the Earth's sidereal rotation period. This synchronisation means that for an observer at a fixed location on Earth, a satellite in a geosynchronous orbit returns to exactly the

APPENDIX B – GNSS GLOSSARY

same place in the sky at exactly the same time each day.

GLONASS (Globalnaya Navigazionnaya Sputnikovaya Sistema)
Russia's global navigation satellite system.

GPS (Global Positioning System)
Full name is NAVSTAR Global Positioning System. The global navigation satellite system operated by the United States of America.

Great Circle
The shortest distance between any two points along the surface of a sphere or ellipsoid, and therefore the shortest navigation distance between any two points on the Earth. Also called Geodesic Line.

Heading
The direction in which a vessel points or heads at any instant, expressed in degrees 000° clockwise through 360° and may be referenced to true north, magnetic north, or grid north. The heading of a vessel is also called the ship's head. Heading is a constantly changing value as the vessel oscillates or yaws across the course due to the effects of the air or sea, cross currents and steering errors.

Horizontal Dilution of Precision (HDOP)
[See *Dilution of Precision (DOP)*]

Horizontal and Time Dilution of Precision (HTDOP)
[See *Dilution of Precision (DOP)*]

Integer Ambiguity Estimates
Carrier phase ambiguity estimates which are only allowed to take on integer values.

Iono-Free Carrier Phase Observation
A linear combination of L1 and L2 carrier phase measurements which provides an estimate of the carrier phase observation on one frequency with the effects of the ionosphere removed. It provides a different ambiguity value (non-integer) than a simple measurement on that frequency.

Kinematic
The user's GNSS antenna is moving. In GNSS, this term is typically used with precise carrier phase positioning and the term dynamic is used with pseudorange positioning.

L-Band
L-Band is a frequency range between 390 MHz and 1.55 GHz which is used for satellite communications and for terrestrial communications between satellite equipment. L-Band includes the GNSS carrier frequencies L1, L2, L5 and several Precise Point Positioning service providers satellite broadcast signals.

Lane
A particular discrete ambiguity value on one carrier phase range measurement or double-difference carrier phase observation. The type of measurement is not specified (L1, L2, L1-L2, iono-free).

Magnetic Bearing
Bearing relative to magnetic north; compass bearing corrected for deviation.

Magnetic Heading
Heading relative to magnetic north.

Magnetic Variation
The angle between the magnetic and geographic meridians at any place, expressed in degrees and minutes east or west to indicate the direction of magnetic north from true north.

Mask Angle
The minimum GNSS satellite elevation angle permitted by a particular receiver design. Satellites below this angle will not be used in position solution.

Misclosure
The gap between a receiver's predicted and actual position.

Moving Base Station
The GNSS receiver which is acting as the reference point but is in motion. It has an estimated position and transmits messages for the rover receiver to use to calculate its position.

Multipath Errors
GNSS positioning errors caused by the interaction of the satellite signal and its reflections.

Nanosecond
One billionth of a second (1×10^{-9} second).

Narrow Lane
The GPS observable obtained by summing the carrier phase observations on the L1 and L2 frequencies. The narrow lane observable can help resolve carrier-phase ambiguities.

Network RTK
With Network RTK, corrections are generated

APPENDIX B – GNSS GLOSSARY

from a base station network instead of from a single base station. These corrections can remove more spatially correlated errors and thus improve the RTK performance as opposed to the traditional RTK. Network RTK uses permanent base station installations, allowing kinematic GNSS users to achieve centimetre (sub-inch) accuracies without the need for setting up a GNSS base station on a known site.

Observation
Any measurement provided by the receiver.

Observation Set
A set of receiver measurements, taken at a given time, that includes one time for all measurements, and the following for each satellite tracked: PRN code, pseudorange or carrier phase or both, lock time count, signal strength and tracking status.

Parity
The even or odd quality of the number of ones or zeroes in a binary code. Parity is often used to determine the integrity of data especially after transmission.

P-Code
Precise code or protected code. A pseudorandom string of bits that is used by GPS receivers to determine the range to the transmitting GPS satellite. P-code is replaced by an encrypted Y-code when Anti-Spoofing is active. Y-code is intended to be available only to authorised (primarily military) users. [See *Anti-Spoofing, (C/A) Code and Y-Code*]

PDOP
Position Dilution of Precision [See *Dilution of Precision (DOP)*]

Post-Processing
A processing mode in which a base station is placed at a known reference point and a rover is used for gathering positions. Accurate coordinates are generated by taking data stored from the receivers and processing them using post-processing software.

Precise Positioning Service (PPS)
The GNSS positioning, velocity and time service which is available on a continuous, worldwide basis to users authorised by the U.S. Department of Defense (typically using P-code).

Pseudorandom Noise Number
A number assigned by the GNSS system designers to a given set of pseudorandom codes. Typically, a particular satellite will keep its PRN (and hence its code assignment) indefinitely, or at least for a long period of time. It is commonly used as a way to label a particular satellite.

Pseudolite
An Earth-based transmitter designed to mimic a satellite.

Pseudorange
The calculated range from the GNSS receiver to the satellite determined by taking the difference between the measured satellite transmit time and the receiver time of measurement, and multiplying by the speed of light. Contains several sources of error.

Pseudorange Measurements
Measurements made using one of the pseudorandom codes on the GNSS signals. They provide an unambiguous measure of the range to the satellite including the effect of the satellite and user clock biases.

Radio Technical Commission for Aeronautics (RTCA)
An organisation which developed and defined a message format for differential positioning.

Radio Technical Commission for Maritime Services (RTCM)
An organisation which developed and defined the SC-104 message format for differential positioning.

Real-Time Kinematic (RTK)
A technique to improve GNSS accuracy that uses pseudorange errors calculated by a base station at a known location to improve the measurements made by other GNSS receivers within the same general geographic area. This type of differential positioning is based on observations of the carrier phase.

Receiver Channels
A GNSS receiver specification which indicates the number of independent hardware signal processing channels included in the receiver design.

Reference Satellite
In a double-difference implementation, measurements are differenced between different satellites on one receiver in order

APPENDIX B – GNSS GLOSSARY

to cancel the correlated errors. Usually one satellite is chosen as the "reference," and all others are differenced with it.

Reference Station
[See *Base Station*]

Relative Bearing
Bearing relative to heading or to the vessel.

Remote Station
[See *Rover Station*]

Residual
In the context of measurement, the residual is the difference between the measurement predicted by the computed solution and the actual measurement.

Route
A planned course of travel, usually composed of more than one navigation leg.

Rover Station
The GNSS receiver which does not know its position and needs to receive measurements from a base station to calculate differential GNSS positions. (The terms remote and rover are interchangeable.)

Satellite-Based Augmentation System (SBAS)
A type of geostationary satellite system that improves the accuracy, integrity and availability of the basic GNSS signals across a large geographical region. This includes WAAS, EGNOS, BDSBAS, MSAS, GAGAN and SDCM.

Selective Availability (SA)
The method used in the past by the U.S. Department of Defense to control access to the full accuracy achievable by civilian GPS equipment. SA was disabled in May 2000.

Sidereal Day
A sidereal day is the rotation period of the Earth relative to the equinox and is equal to one calendar day (the mean solar day) minus approximately four minutes.

Spheroid
Sometimes known as ellipsoid; a perfect mathematical figure which very closely approximates the geoid. Used as a surface of reference for geodetic surveys.

Standard Positioning Service (SPS)
A positioning service made available by the U.S. Department of Defense which is available to all GPS civilian users on a continuous, worldwide basis (typically using C/A Code).

Space Vehicle ID (SV)
Sometimes used as SVID. A unique number assigned to each satellite for identification purposes. The 'space vehicle' is a GNSS satellite.

TDOP
Time Dilution of Precision [See *Dilution of Precision (DOP)*]

Time-To-First-Fix (TTFF)
The actual time required by a GNSS receiver to achieve a position solution. This specification will vary with the operating state of the receiver, the length of time since the last position fix, the location of the last fix, and the specific receiver design.

Track Made Good
The single resultant direction from a point of departure to a point of arrival or subsequent position at any given time; may be considered synonymous with Course Made Good.

True Bearing
Bearing relative to true north; compass bearing corrected for compass error.

True Heading
Heading relative to true north.

Undulation
The distance of the geoid above (positive) or below (negative) the mathematical reference ellipsoid (spheroid). Also known as geoidal separation, geoidal undulation, geoidal height.

Update Rate
The GNSS receiver specification which indicates the solution rate provided by the receiver when operating normally.

UTC
[See *Coordinated Universal Time*]

VDOP
Vertical Dilution of Precision [See *Dilution of Precision (DOP)*]

Waypoint
A reference point on a track.

Wideband Antenna
A GNSS antenna that is capable of receiving multiple Global Navigation Satellite Systems (GNSS) including GPS, GLONASS, BeiDou and Galileo frequencies.

Wide Lane
A particular integer ambiguity value on one

APPENDIX C – STANDARDS AND REFERENCES

This appendix provides links to companies and agencies engaged in activities related to GNSS. Web site addresses are subject to change; however, they are accurate at the time of this book's publication.

Hexagon Autonomy & Positioning

Contact your local dealer first for more information. To locate a dealer in your area or to resolve a technical problem:

Phone:

U.S. & Canada: 1-800-668-2835

China: 0086-21-68882300

Europe: 44-1993-848-736

SE Asia and Australia: 61-400-883-601

E-mail: sales-inside.ap@hexagon.com

Website: hexagon.com/autonomy-and-positioning

Other sources of information about GNSS and autonomy:

Arinc: arinc.com

BeiDou Navigation Satellite System: en.beidou.gov.cn

European Space Agency (Galileo and EGNOS information): esa.int/Our_Activities/Navigation

European Union Agency for the Space Programme: euspa.europa.eu

Geodetic Survey of Canada: geod.nrcan.gc.ca

GPS System: gps.gov

Hexagon | NovAtel: novatel.com

Hexagon | AutonomouStuff: autonomoustuff.com

Hexagon | Veripos: veripos.com

Indian Space Research Organisation (NavIC information): isro.org/index.aspx

National Geodetic Survey: ngs.noaa.gov

National Marine Electronics Association (NMEA): nmea.org

NAVSTAR GPS Operations: tycho.usno.navy.mil/gpsinfo.html

Quasi-Zenith Satellite System (QZSS): qzs.jp/en

Radio Technical Commission for Aeronautics (RTCA): rtca.org

Radio Technical Commission for Maritime Services (RTCM): rtcm.org

Russian Federal Space Agency (GLONASS information): hglonass-iac.ru/en

Society of Automotive Engineers (SAE): sae.org

APPENDIX D – ACKNOWLEDGEMENTS

Hexagon's Autonomy & Positioning division would like to thank the following individuals for their contributions to the development of this book.

Charles Jeffrey	B.Sc.EE Engineering Physics
Kurt Bahan	B. Sc., Geomatics Engineering, P. Eng
Michael Olynik	B. Sc., M. Sc., Geomatics Engineering, P. Eng.
Todd Richert	B. Sc., M. Sc. Geomatics Engineering, P. Eng., BEd.
Maged Shenouda	B. Sc., M. Sc. Electrical Engineering, P. Eng.
Anil Sinha	B. Sc., Geomatics Engineering, P. Eng
Peter Soar	B. Sc. Engineering, former Royal Artillery Officer
Kevin Vance	B. Sc., Geomatics Engineering, P. Eng
Neil Gerein	B. Sc., M. Sc. Electrical Engineering
Paul Verlaine Gakne	B. Sc. M. Sc. Telecommunications, PhD. Geomatics Engineering
David Bradley	M. A. English Language and Literature
Jan Diep	B. Sc. Engineering, P. Eng.
Bryan Leedham	B. Sc. Geomatics Engineering, P. Eng.
Roger Munro	Dipl. Electronic Engineering Technology
Jessica Tedford, CD	B. Comm. Journalism
Lee Baldwin	B. B.A. Management Information Systems
Maria Meijer	B. Sc., M. Sc. Genetics and Cell Biology
Caitrin Junker	B. Mgt. International Business
Stig Pedersen	B. Sc., M. Sc. Business Administration
Ryan Dixon	B. Sc. Geomatics Engineering, P. Eng.
Brett Kruger	B. A.Sc., M. A.Sc. Electrical Engineering, P. Eng.
Caleb Imig	B. Sc. Mechanical Engineering
John Fleming	HND, B. Sc. Applied Physics and Microelectronics
Zoltan Molnar	M. Sc. Electrical Engineering, P. Eng.
Gord Heidinger	B. Sc. Electrical Engineering

www.ingramcontent.com/pod-product-compliance
Lightning Source LLC
Chambersburg PA
CBHW042050290426
44110CB00001B/14